U.S. Department of Transportation
National Highway Traffic Safety Administration

DOT HS 811 572 June 2012

An Analysis of Recent Improvements to Vehicle Safety

DISCLAIMER

This publication is distributed by the U.S. Department of Transportation, National Highway Traffic Safety Administration, in the interest of information exchange. The opinions, findings, and conclusions expressed in this publication are those of the authors and not necessarily those of the Department of Transportation or the National Highway Traffic Safety Administration. The United States Government assumes no liability for its contents or use thereof. If trade names, manufacturers' names, or specific products are mentioned, it is because they are considered essential to the object of the publication and should not be construed as an endorsement. The United States Government does not endorse products or manufacturers.

Technical Report Documentation Page

1. Report No. DOT HS 811 572	2. Government Accession No.	3. Recipient's Catalog No.
4. Title and Subtitle An Analysis of Recent Improvements to Vehicle Safety		5. Report Date June 2012
		6. Performing Organization Code NVS-421
7. Author(s) Glassbrenner, Donna, Ph.D.		8. Performing Organization Report No.
9. Performing Organization Name and Address Mathematical Analysis Division, National Center for Statistics and Analysis National Highway Traffic Safety Administration U.S. Department of Transportation, NVS-421 1200 New Jersey Avenue SE. Washington, DC 20590		10. Work Unit No. (TRAIS)
		11. Contract or Grant No.
12. Sponsoring Agency Name and Address Mathematical Analysis Division, National Center for Statistics and Analysis National Highway Traffic Safety Administration U.S. Department of Transportation, NVS-421 1200 New Jersey Avenue SE. Washington, DC 20590		13. Type of Report and Period Covered NHTSA Technical Report
		14. Sponsoring Agency Code
15. Supplementary Notes		

Abstract

This report seeks to quantify the improved safety of newer vehicles and its contribution to historically low fatality and injury rates experienced in the United States in recent years. We develop statistical models of crash avoidance and crashworthiness and apply the results of these models to estimate the likely result of replacing newer vehicles with older vehicles and vice versa, while controlling for human and environmental factors that would otherwise cloud the result.

The analysis finds remarkable improvements to vehicle safety. We estimate that the likelihood of crashing in 100,000 miles of driving has decreased from 30 percent in a model year 2000 car to 25 percent in a model year 2008 one, when both vehicles are driven "as new". The likelihood of escaping a crash uninjured has improved from 79 to 82 percent as a result of improvements between the 2000 and 2008 car fleets. Improvements are also found for light trucks and vans, and for the chances of surviving a crash and avoiding incapacitation.

The nationwide impact of these advancements is substantial. We estimate that improvements made after the model year 2000 fleet prevented the crashes of 700,000 vehicles; prevented or mitigated the injuries of 1 million occupants; and saved 2,000 lives in the 2008 calendar year alone. Of the 9 million passenger vehicles that were in crashes, the crashes of an estimated 200,000 of them were preventable by improvements to the model year 2008 fleet, and the injuries of 300,000 of their 12 million occupants would have been prevented or mitigated, including saving 600 lives.

17. Key Words Traffic safety, vehicle safety, crashworthiness, crash avoidance, fatalities, injuries		18. Distribution Statement This document is available to the public from the National Technical Information Service www.ntis.gov	
19. Security Classif. (of this report) Unclassified	20. Security Classif. (of this page) Unclassified	21. No. of Pages 85	22. Price

Form DOT F 1700.7 (8-72)   Reproduction of completed page authorized

TABLE of CONTENTS

1. Summary 1

2. Objective 2

3. Some Preliminaries 4

    3.1. Definition of Crash Avoidance 4
    3.2. Definition of Crashworthiness 5
    3.3. Travel Worthiness 6
    3.4. Data Sources 6
    3.5. Treatment of Missing Data 7

4. Raw Estimates 8

    4.1. Crash Avoidance 8
    4.2. Crashworthiness 10

5. Model Fitting 15

    5.1. The Crash Avoidance Model 15
    5.2. The Crashworthiness Model 25
    5.3. Limitations of Both Models 39

6. How Much Safer Are Newer Vehicles? 40

    6.1. The Reduced Likelihood of Crashing 40
    6.2. The Reduced Likelihood of Injury in a Crash 44

7. Reductions in Crashes, Injuries, and Fatalities 50

    7.1. Crashes Avoided and Avoidable 51
    7.2. Injuries Mitigated and Mitigatable 57
    7.3. Lives Saved and Savable 63
    7.4. The Benefits of Modeling in Assessing Fleet Improvements 64

8. References 70

9. Appendix 71

    9.1. Raw Crashworthiness Estimates 71
    9.2. Parameter Estimates for the Crash Avoidance Model 77
    9.3. Parameter Estimates for the Crashworthiness Model 78

1. Summary

The United States has achieved historically low fatality and injury rates in recent years (NHTSA's National Center for Statistics and Analysis, 2009). This report seeks to understand the role that improved vehicle safety played in this achievement. In theory, vehicle safety improvements could be initiated by various sources (government, vehicle manufacturers, or others) and could be mandatory or voluntary in nature. Without worrying about their nature or source, one should be able to assess their collective impact by examining crash databases (at least in combination with information on exposure, such as miles driven). This is the approach we take in our analysis. Namely, we assess the crashworthiness and crash avoidance performance of vehicles from different model year fleets, as evidenced by the crashes that occurred and miles driven, and translate the difference in performance into crash and injury outcomes.

Along the way, we address basic questions about vehicle safety, such as: How should we quantify the crash avoidance capacity of a vehicle? What about its crashworthiness? How likely am I to crash in, say, 100,000 miles of driving? What is my chance of injury if I do crash? Are there confounding factors that should be taken into account in order to elicit correct assessments of these quantities?

Indeed we will find confounding factors, and so we conduct our analysis via statistical modeling in order to control for factors that would otherwise cloud our assessment. We assess whether the forms of our models are appropriate, whether they appear to successfully control for the factors we designed them to control for, and what results one would obtain under alternative approaches, including one that is not model-based. In the end, these appear to support our chosen methodology.

Our study finds remarkable improvements to vehicle safety. We estimate that the likelihood of crashing in 100,000 miles of driving has decreased from 30 percent in a model year 2000 car to 25 percent in a model year 2008 one, when both vehicles are driven "as new". The likelihood of escaping a crash uninjured has improved from 79 to 82 percent as a result of improvements between the 2000 and 2008 car fleets. Improvements are also found for light trucks and vans, and for the chances of surviving a crash and avoiding incapacitation.

The nationwide impact of these advancements is substantial. We estimate that improvements made after the model year 2000 fleet prevented the crashes of 700,000 vehicles; prevented or mitigated the injuries of 1 million occupants; and saved 2,000 lives in the 2008 calendar year alone. Of the 9 million passenger vehicles that were in crashes, the crashes of an estimated 200,000 of them were preventable by improvements to the model year 2008 fleet, and the injuries of 300,000 of their 12 million occupants would have been prevented or mitigated, including saving 600 lives.

2. Objective

The objective in this study is to isolate and quantify the vehicle component in recent improvements to traffic safety from human and environmental effects.

Traveling by vehicle has gotten remarkably safer in recent years. Fatality and injury rates reached new lows in 2009, with 1.14 people killed and 75 people injured per 100 million vehicle miles, compared to 1.55 fatalities and 120 injured people 10 years ago (NHTSA's National Center for Statistics and Analysis, 2010).

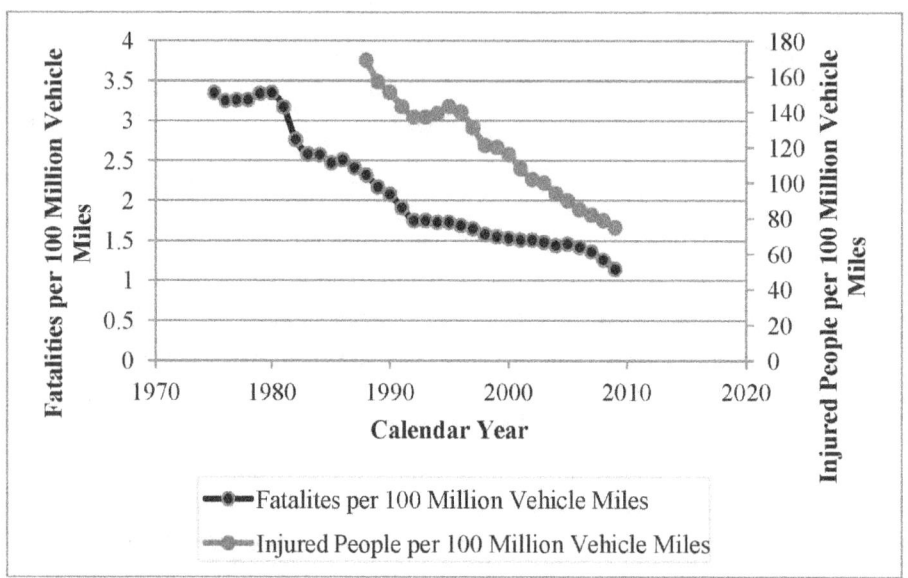

Figure 1-1: Fatalities and People Injured per 100 Million Miles Traveled

Speaking broadly, traffic safety is influenced by three components: human factors, vehicle/equipment factors, and environmental factors, as illustrated in Figure 1-2.

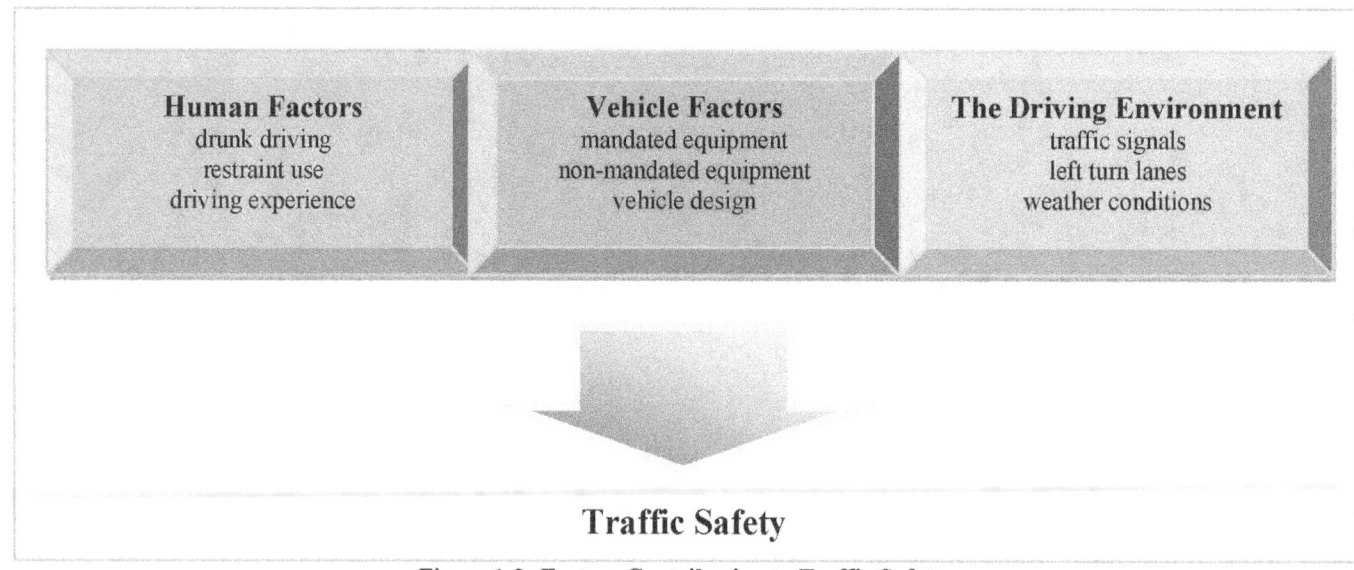

Figure 1-2: Factors Contributing to Traffic Safety

As each of these factors contributes to traffic safety, each may have contributed to the recent improvements in it. Our objective in this study is to isolate the vehicle/equipment component. That is, in broad terms, we ask "How much safer are newer vehicles?"

We limit our study to passenger vehicles, that is, to cars, pickup trucks, vans, and sport utility vehicles with gross vehicle weight ratings of 10,000 pounds or less. These vehicles are involved in most of the crashes,[1] and most vehicle regulations promulgated by the National Highway Traffic Safety Administration (NHTSA) are directed towards them.[2] In addition, as we are interested in *recent* traffic safety improvements, we shall primarily focus on improvements since the 2000 model year.

Our study seeks to answer fundamental questions, such as: How likely am I to crash in, say, 100,000 miles of driving a model year 2000 car, and by how much has this reduced (if at all) for a model year 2008 car? If I am in a crash, how much better will I fare in a model year 2008 car than in a model year 2000 one? By how much has my chance of surviving the crash increased? What about my chance of walking away uninjured? Extrapolating to their nationwide implications, how many crashes were prevented and injuries mitigated by these improved outcomes?

A crucial issue we expressly do *not* address, or even attempt to address, in this study is to identify the particular sources of the improvements. It will be natural to wonder *what* lead to the improvements we quantify. For instance, as we quantify the reduced likelihood of crashing in a newer vehicle (controlling for human and environmental factors), we will wonder which particular technologies and innovations in vehicle design were the primary agents in this collective societal benefit. We may also wonder whether the benefits chiefly stemmed from regulatory requirements, government safety ratings,[3] or by independent innovations undertaken by vehicle and equipment manufacturers. While these are important questions, investigating them would constitute a major undertaking and is beyond the scope of this study.

Our study is in the spirit of the landmark assessment (Kahane, 2004) which assessed the combined safety improvement of all Federal Motor Vehicle Safety Standards (FMVSS) issued by NHTSA, as well as technologies not mandated or regulated under the FMVSS. That is, while we seek to quantify the combined benefit of *recent* changes to vehicles, Kahane assessed the combined benefit of changes since 1960.[4] Phrased differently, as we ask how much safer a model year 2008 car is than a model year 2000, for example, Kahane asked how much safer a contemporary vehicle was than a model year 1960 vehicle.[5]

Our analysis attempts to control for non-vehicle factors through the use of statistical modeling. As such, it is important to consider the degree to which our attempts at control were successful when deciding whether to believe our estimates of reduced crash likelihoods, improved injury outcomes, and the collective societal benefit. To this end, we present and discuss the relative merits of alternative calculations that are made with no statistical models or models applied differently, in addition to discussing the extent to which our chosen methodology controls for confounders.

---

[1] In 2008, 94 percent of the vehicles involved in crashes were passenger vehicles (NHTSA's National Center for Statistics and Analysis, 2008).
[2] A relatively minor point is that we also exclude pre-production vehicles, as these are generally not available for sale to the public.
[3] In the New Car Assessment Program (NCAP), NHTSA issues "five-star" safety ratings and identifies vehicles with recommended safety technologies, to inform consumers in their vehicle purchase.
[4] Kahane found that in total, regulatory requirements saved 280,000 lives between January 1, 1960, and December 31, 2002, with voluntary improvements by vehicle and equipment manufacturers saving an additional 44,000 lives.
[5] A key difference between our assessments lies in our methods. Kahane starts with estimated effectiveness of individual technologies to build the collective improvement. In our study, we extrapolate the collective safety improvement by filtering our human and environmental factors from crash data. That is, Kahane builds from within the vehicle/safety component in Figure 1-2, while we filter out non-vehicle components. Each method exercises control for human and environmental factors, and so we would expect the two methodological approaches to produce comparable results. Indeed our assessment is consistent with Kahane's results.

3. Some Preliminaries

We shall analyze the improved safety of newer vehicles in terms of improvements to their crashworthiness and their capacities to avoid crashes. This chapter presents formal definitions of these concepts and a concept, travel worthiness, that reflects an overall measure of safety. Our definitions will utilize the KABCO injury scale,[6] but could just as well use any ordinal injury classification system. We also discuss the data sources we will use to estimate these quantities and our treatment of missing values in these data.

3.1 Definition of Crash Avoidance

Definition Given a type of crashes c and vehicles v, we define the *crash avoidance* (*capacity*) to be the probability that a vehicle of the given type driven for 100,000 miles does not get into any crashes of the given type, i.e.:

$$CA = P(\text{no crashes of type } c \mid \text{a vehicle of type } v \text{ travels 100,000 miles})$$

That is, we have a probabilistic experiment, which we could refer to as the *Travel Experiment*, in which we subject a vehicle[7] of type v to 100,000 miles of travel[8] and record the number of crashes *Crashes* of type c that the vehicle gets involved in.[9] *Crashes* is a random variable taking nonnegative integer values, and crash avoidance is the probability that *Crashes* takes the value 0.

Assuming the distribution of vehicle crashes[10] over miles driven is negative binomial,[11] crash avoidance is related to the crash rate via

$$CA = (1 + 0.00001\, CR)^{-100,000} \tag{3-1}$$

where CR denotes the crash rate (i.e., the number of crashes of type c in 100,000 miles of driving a vehicle of type v). With crash avoidance defined using such a large number of miles (100,000), CA is also approximately equal to the value it would have if we assumed crashes were Poisson-distributed, namely e^{-CR}, where e denotes the base of the natural logarithm. (In the vehicle and crash types we consider, the difference will be at most 0.000001. Figure 3-1 displays both as functions of the crash rate, and the two are indistinguishable for crash rates under two vehicle involvements per 100,000 miles. In this figure, the chart in the right panel is a magnification of that from the left.)[12][13]

---

[6] See (NHTSA NASS, 2008) for information on the KABCO system. As mentioned previously, the system classifies: K = fatal injury, A = incapacitating injury, B = non-incapacitating injury, C = possible injury, O = no injury.

[7] Although the example of a vehicle type (model year 2000 cars with a sober driver) that we used to illustrate crashworthiness incorporated driver sobriety, we won't have sufficient information to quantify crash avoidance estimates that incorporate driver sobriety. The vehicle types we use for estimates of crash avoidance will be more broad (e.g., model year 2000 cars). However the *notion* of crash avoidance applies in theory to any defining characteristics of vehicles to which our Travel Experiment could be applied.

[8] Using 100,000 miles as a reference exposure level will put our crash rates and crash avoidance estimates in ranges that are easier to interpret. With the average passenger vehicle logging about 12,000 miles per year (3 trillion miles traveled by 240 million passenger vehicles in 2008), 100,000 miles is the approximate travel logged by a vehicle in 8 years' time.

[9] Obviously the number of crashes a vehicle gets in depends on who is driving the vehicle (a sober, experienced driver, etc.) and under what conditions (icy or clear conditions, highway driving, etc.). Our notation suppresses information on the driving conditions, which will be clear from context when we apply our definition.

[10] Our unit of measure here is vehicle crashes (i.e., the number of vehicles in crashes). For instance, a head-on collision of two vehicles comprises two vehicle crashes. We will sometimes refer to vehicle crashes as vehicle involvements or simply as "crashes" (e.g., saying "crash rate" instead of "vehicle crash rate"), and rely on context to determine the precise unit intended.

[11] The negative binomial and Poisson distributions are commonly used in modeling crashes over time or miles. See, e.g., (Poch & Mannering, 1996), (Turner-Fairbank Highway Research Center, 2005). The negative binomial is a more general family of distributions, with the Poisson distribution for a given crash rate λ arising as a limiting value of the negative binomial distribution, namely $\text{Poisson}(\lambda) = \lim_{r \to \infty} NB(r, \lambda/(\lambda+r))$. The (discrete) negative binomial distribution $NB(r,p)$ is defined as $P(X=k) = (k+r-1)!\,(1-p)^r\, p^k/(k!\,(r-1)!)$ for $k = 0,1,2,\ldots$. We are modeling the number of crashes in 100,000 miles of travel for a given crash and vehicle type as $NB(100{,}000,\ CR/(100{,}000+CR))$, where CR is the average number of crashes of the given type experienced by a vehicle of the given type in 100,000 miles of travel. Thus, $CA = P(X=0) = (1 + 0.00001\,CR)^{-100{,}000}$.

[12] With $e^{-x} = \lim (1+x/n)^{-n}$, the limit going from 1 to infinity, we note that $e^{-x} \approx (1+x/n)^{-n}$ for large n. (Rudin, 1976)

[13] We need to assume a distribution for crashes over miles traveled in order to estimate crash avoidance. Although we have data on miles driven and crashes, we don't know which vehicles in our crash database went 100,000 miles without crashing. However, we can estimate the crash rate, and estimate crash avoidance via Equation (3-1). As noted earlier with the Poisson distribution, modeling

Crashes being relatively rare,[14] our estimated crash avoidance is approximately equal[15] to one minus the crash rate. That is, $CA \approx 1 - CR$.

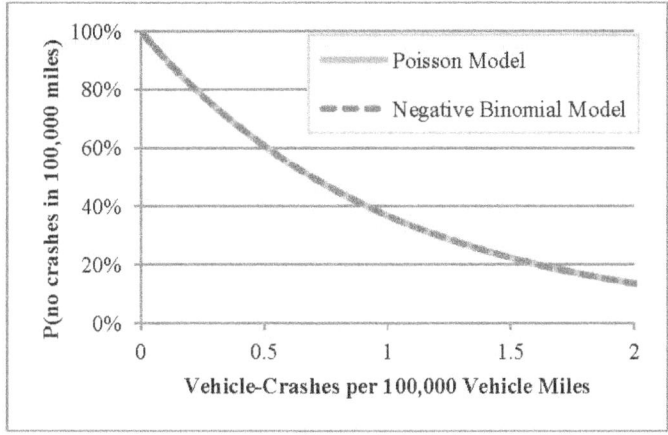 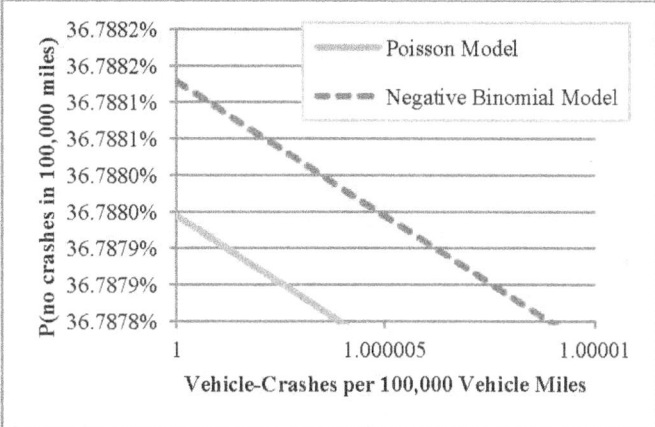

Figure 3-1: Poisson and Negative Binomial Models of Crash Avoidance

3.2 Definition of Crashworthiness

Definition Given a type c of crashes (e.g., frontal crashes), a type v of vehicles (e.g., model year 2000 cars with a sober driver), a type o of occupants (e.g., belted 25- to 65-year-old women), and an injury threshold z (e.g., non-incapacitating injuries), we define *crashworthiness* to be the probability that an occupant of the given type in a vehicle of the given type in a crash of the given type sustains an injury no worse than the given threshold, i.e.:

$$CW = P(Injury \leq z \mid \text{an occupant of type } o \text{ is in a crash of type } c \text{ in a vehicle of type } v)$$

More formally, we have a probabilistic experiment, which we could refer to as the *Crash Experiment*, in which we subject an occupant of type o in a vehicle of type v to a crash of type c and record the level of injury *Injury* sustained by the occupant. Our injury scales will be discrete (and ordinal[16]), and so *Injury* is a discrete random variable. For improved readability, our notation CW suppresses the dependence on c, v, o, and z.

We can extend our notion of crashworthiness to one that is independent of crash type by specifying a distribution of crash types. Our study defines crashworthiness using five[17] crash types – frontal, near side, far side, rollover, and other. Given a distribution $(p_1, \ldots p_5)$ of their relative incidence (e.g., over a given time period such as in the calendar year 2008), we can imagine an experiment in which

---

crashes using a distribution other than the negative binomial could yield different crash avoidance estimates from the same data (although they are similar in our case).

[14] In 2008, there were 9 million passenger vehicles in crashes, and these vehicles traveled a collective 3 trillion miles, for a crash rate of 0.34 crashes per 100,000 miles driven.

[15] Were crash rates high, the crash rate would be very different from the chance of crashing at least once in 100,000 miles. For instance, if we crashed every other mile on average, the former would be 0.5, while the latter would be nearly 1 (i.e., nearly 100%). However, with low crash rates, the two are guaranteed to be at least somewhat close: The Taylor series for $(1+0.00001x)^{-100,000}$ centered at $x=0$ is $1 - x + 0.500005\, x^2 - 0.17\, x^3 + \ldots$, so $|(1+0.00001x)^{-100,000} - (1-x)| < 0.500005\, |x|^2$. Our crash rates will at most be 0.35, and we know a priori that the difference between CA and $1 - CR$ will be at most $(0.500005)(0.35)^2 \approx 0.06$. That is, approximating crash avoidance as one minus the crash rate would be accurate within six percentage points.

[16] Although the KABCO classification contains codes for unknown injuries ("I" for injuries of unknown severity, "U" for persons for whom it is not known whether they are injured), we assume for the purpose of defining crashworthiness that all injuries can be classified with known values in the Crash Experiment. When we estimate crashworthiness from crash data, we will impute unknown KABCO values.

[17] Crash avoidance used fewer crash types (frontal, rollover, side, and other) because it is defined at the vehicle level. We have defined crashworthiness at the occupant level to reflect that the ability of a vehicle to protect occupants will be different for e.g. belted versus unbelted occupants, and may well depend on characteristics such as whether the occupant is elderly or female. We could define a vehicle-level notion of crashworthiness by specifying a distribution of occupant types and the relative occurrence of near side versus far side crashes, but feel it would be somewhat disingenuous to do so, as we will discuss in Section 6.2.

crashes of various types are generated according to the given distribution, followed by performing the Crash Experiment. Our crash type- independent crashworthiness is then:

$$P(Injury \leq z \mid \text{an occupant of type } o \text{ is in a crash in a vehicle of type } v) = \sum_{k=1}^{5} p_k CW_k$$

where the $CW_1, \ldots CW_5$ are as defined in the preceding definition. We shall refer to both the crash type specific and the generalized notion as crashworthiness and rely on context to determine which is intended.

3.3 Travel Worthiness

Our notions of crashworthiness and crash avoidance could be combined into a single measure of vehicle safety. Namely, one could define the travel worthiness of a vehicle of a given type to be the probability that an occupant of a given type would sustain at worst a given level of injury in 100,000 miles of travel. That is, the underlying probability experiment combines the experiments Travel and Crash to subject an occupant of the given type to 100,000 miles of travel in a vehicle of the given type and records whether the occupant sustained any injuries beyond the given threshold at any time during the travels. Although we find the idea interesting and think that the statistical models for crash avoidance and crashworthiness developed in this paper could be used to assess fleet improvements in travel worthiness, such is beyond the scope of this paper.

3.4 Data Sources

We shall estimate and model crash avoidance and crashworthiness using crash data from NHTSA's Fatality Analysis Reporting System (FARS) and General Estimates System (GES), mileage data from the U.S. Department of Transportation's National Household Travel Survey and the Federal Highway Administration, and vehicle registrations from R.L. Polk and Company's National Vehicle Population Profile.

The Fatality Analysis Reporting System (FARS)
FARS is an annual census of motor vehicle crashes in the United States, excluding U.S. territories other than Puerto Rico, occurring on trafficways[18] in which at least one person (whether a motorist, pedestrian, or pedalcyclist) involved in the crash dies within 30 days of the crash. FARS contains a variety of information concerning the vehicles and persons involved in the crash, as compiled from police accident reports, death certificates, and other sources.[19]

The General Estimates System (GES)
The GES is a data system containing information from a probability sample of police-reported[20] crashes occurring on U.S. trafficways, excluding U.S. territories. These crashes may have resulted in fatalities, injuries, and/or property damage. GES contains a variety of information concerning the vehicles and persons involved in the crash sample, as compiled from police accident reports.[21]

The National Household Travel Survey
The NHTS is a nationwide survey conducted on a probability sample of U.S. households, which report data on the trips they conduct during the study period, and various demographic characteristics. The sample excludes residents of college dormitories, nursing homes, medical institutions, prisons, and military bases. We use data from the 2001 NHTS, which collected data from 66,000 households and was conducted between April 2001 and May 2002.

Vehicle Miles Traveled by Vehicle Type
The Federal Highway Administration (FHWA) compiles data on the collective vehicle miles traveled by vehicle type submitted by States each year into nationwide estimates. The estimates are published in the FHWA *Highway Statistics* series (FHWA, 2008).

---

[18] Trafficways include road shoulders, sidewalks, medians, and roadsides up to the property line. For this and other crash terminology (such as the definition of "motor vehicle"), FARS and GES use the American National Standard definitions, set by the National Safety Council and documented in (D16 Committee on Classification of Motor Vehicle Traffic Accidents, 2007).
[19] While FARS contains information on crashes in Puerto Rico, we do not include these crashes in our analyses because there is no corresponding information in GES. For more information on FARS, see (Chang, 2009).
[20] We note that the standards under which a crash must be reported to the police differ by State, and so there is some degree of non-uniformity in the GES data. We also note that not all crashes that should be reported to the police are, and so GES under-represents to some degree crashes that meet the thresholds for police-reporting (i.e. would underestimate their number).
[21] For more information on GES, see (National Highway Traffic Safety Administration, 2008).

The National Vehicle Population Profile

R.L. Polk's National Vehicle Population Profile provides an annual census of passenger cars and light trucks and vans (LTVs)[22] registered for on-road operation in the United States, including Puerto Rico but excluding other U.S. territories. The Polk files contain a variety of information concerning the vehicles registered, as compiled from State Department of Motor Vehicles offices.[23]

3.5 Treatment of Missing Data

Our databases have unknown values, and FARS provides multiple model-based imputations of driver alcohol. We impute unknown values for other FARS and GES variables with five hotdeck imputations, using the following donor cells.

Table 3-1: Imputation Cells

Variable Imputed	Variables Defining the Imputation Cells
Whether the vehicle has a driver	Vehicle type, crash year
Vehicle type	Crash year
Occupant gender	Vehicle type, crash year
Occupant age category	Vehicle type, crash year
Driver alcohol involvement[24]	Gender, age category
Seating position	Vehicle type, crash year
Injury severity (KABCO)	Crash type, restraint use
Vehicle impact area	Vehicle type, crash year
Vehicle model year	Vehicle type, crash year

These cells are admittedly coarse. It is beyond the objective of this paper to develop sophisticated imputation models.

We use known and unknown values in forming the imputation cells. For instance, one of the imputation cells used to impute occupant gender is unknown vehicle types in crashes in the year 2000.

We chose these imputation cells based on intuitive relationships with the imputed variable. We shall compute imputation errors in our statistical models. However, we recognize that our computed imputation errors only reflect the impact of different implementations of our chosen imputation method and imputation cells. We do not examine the impact of alternative imputation methods or alternative choices of imputation cells.

We derive an imputed crash type from the imputed seating position and vehicle impact variables.

We are less familiar with NHTS and Polk data and do not impute their unknowns. We only use these data in benchmarked estimates, and so we will not (superfluously) distribute these unknown values.

---

[22] LTVs comprise sport utility vehicles, vans, and pickup trucks. For NHTSA's definitions of these terms and the definition of "(passenger) car", see (Chang, 2009).
[23] For more information on the R.L. Polk data files, see www.polk.com.
[24] We only impute alcohol involvement for the GES cases, and use the FARS multiple imputation of alcohol for the FARS cases.

4. Raw Estimates

4.1 Crash Avoidance

Computing crash avoidance is complicated by having relatively limited information on miles traveled. Suppose for instance we wish to compute the estimated probability that a model year 2000 car driven 100,000 miles does not get into any frontal crashes, using data from crashes that occurred in 2008. Using Equation (3-1), we can estimate this quantity as $(1 + 0.00001\, CR)^{-100{,}000}$, where CR denotes the corresponding crash rate. We can estimate the number of frontal crashes of model year 2000 cars that occurred in calendar year 2008 as an average of the five Horvitz-Thompson estimates from the hotdeck imputations of our combined FARS-GES database.[25]

We estimate the denominator of the crash rate (i.e., the collective number of miles driven by model year 2000 cars during 2008) using our NHTS, FHWA, and Polk data. Namely, we fit an exponential model, depicted in Figure 4-1, to the 2001 NHTS estimates of the annual miles driven by a car as a function of its age.[26] [27]

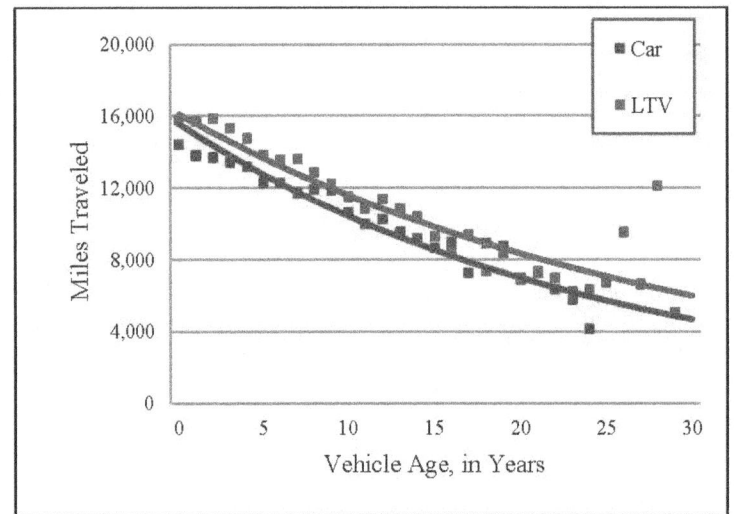

Figure 4-1: Exponential Models of the Miles a Vehicle Travels in a Year

The equations of these models, which are the least-squares fits, are:

$$\log M_{car}(y) = -0.03670349\, y + 9.62234152$$
$$\log M_{LTV}(y) = -0.03404264\, y + 9.69295143 \qquad (4\text{-}1)$$

where *log* denotes the natural logarithm.

We apply registration figures from Polk to estimate the collective vehicle miles driven by model year cars during 2008 and benchmark this to the FHWA estimate of car miles traveled in 2008. In total, the collective number of miles driven by model year 2000 cars during 2008 is estimated as:

---

[25] Namely, we can estimate this quantity as $(C_1 + \ldots + C_5)/5$ where C_i denotes the Horvitz-Thompson estimate of the number of frontal crashes of model year 2000 cars that occurred in calendar year 2008, using the ith hotdeck imputation of our combined FARS-GES database.

[26] We define vehicle age as the difference between the calendar (or crash) year and model year.

[27] NHTSA periodically publishes a report of the miles traveled per vehicle by vehicle age and vehicle type, most recently (Lu, 2006). We did not use the Lu report because it appears to use a different definition of vehicle age. Lu appears to define vehicle age as the age from purchase or manufacture of a vehicle, not the difference between the year of interest and the model year. Also we were not certain whether Lu's classification of vehicle type was Polk's or NHTSA's, which differ in the treatment of crossover utility vehicles.

$$V \frac{M_{car}(8)R_{2000}}{\sum_k M_{car}(2008-k)R_k} \tag{4-2}$$

where V denotes the FHWA estimate of car miles traveled in 2008, k ranges over model year, $M_{car}(.)$ denotes our model from equation (4-1), and R_k denotes the number of model year k cars registered in 2008 from Polk. The index of summation k in the denominator of Equation (4-2) ranges through the model years in the Polk database.[28,29] Figure 4-2 depicts the collective miles traveled in 2008 for each model year and both vehicle types.

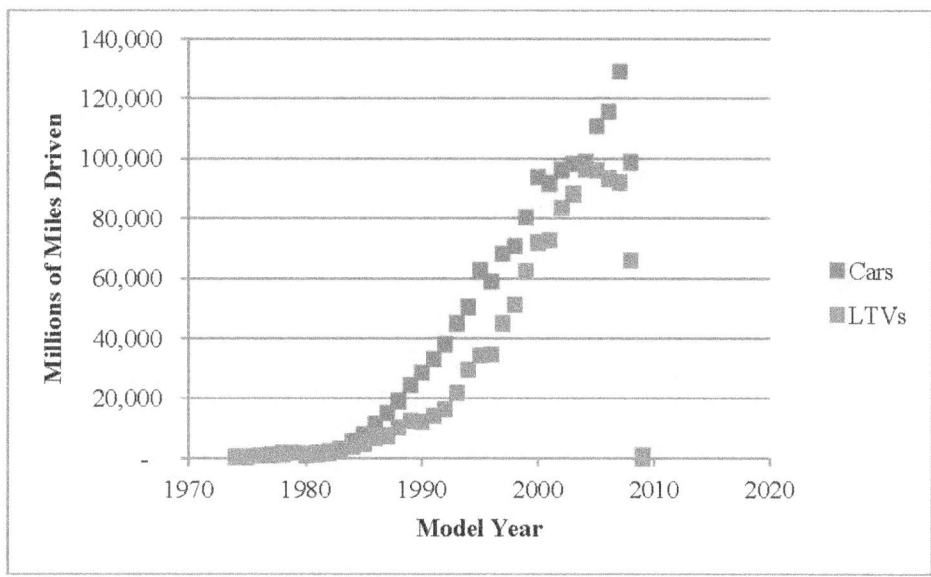

Figure 4-2: Collective Vehicle Miles Traveled by Cars and LTVs in 2008 by Model Year

Returning to our objective of calculating the capacity of model year 2000 cars to avoid frontal crashes based on information from the 2008 calendar year, we have $CA = (1 + 0.00001\ CR)^{-100,000}$, and we now have all of the information needed to perform this calculation: The crash rate CR equals C/M, C being the geometric mean[30] of our five Horvitz-Thompson estimates of the number of frontal crashes of model year 2000 cars that occurred in calendar year 2008 and M being the estimated miles driven by model year 2000 cars in calendar year 2008 (from Equation 4-2).[31,32] Using a similar calculation, one can also compute the likelihood of avoiding crashes of *any* type.[33]

---

[28] The newest vehicles in Polk have an age of negative one (–1) years (e.g., model year 2009 vehicles in 2008). Applying the model from Equation (4-1) would overestimate the miles driven by such vehicles, which typically arrive in dealer showrooms in the last quarter of the calendar year. We estimate the miles driven at age -1 by applying the ratio of the age -1 and age 0 NHTS estimates to the age 0 model predictions. This results in estimates of $M_{car}(-1)=1,897$ and $M_{LTV}(-1)=2,027$.

[29] Polk aggregates registration figures for pre-1974 model year vehicles, and so the earliest model year in the summation in Equation (4-2) will be an aggregated figure. We take $M_{car}(2008-k)$ in this case to be the average of the predicted miles driven by vehicles of age 35-40, which compute to 3,820 miles for cars and 4,528 miles for LTVs.

[30] The geometric mean of values $x_1, ..., x_n$ is $\sqrt[n]{x_1 x_2 \cdots x_n}$.

[31] We have an order-of-operations choice to make here. Other than computing crash avoidance as described, we could compute the crash avoidance in each imputation and average the results (i.e., average$((1+0.00001\ CR_1)^{-100,000}, ..., (1+0.00001\ CR_5)^{-100,000})$, where $CR_1, ..., CR_5$ are the crash rates in the five imputations of our crash dataset). A third way would be to compute crash avoidance from the average crash rate, i.e., $(1+0.00001\ (CR_1+...+CR_5)/5)^{-100,000}$. Had we not developed a crash avoidance model, we likely would have chosen one of these approaches to estimate raw crash avoidance. However our crash avoidance model will model the log crash rate as a linear function of the predictors, and each of our five imputations of the crash data will yield parameter estimates for the linear coefficients. Applying either of the alternative approaches to the model predictions, e.g., $\widehat{CA}=$ average$((1+0.00001\ \widehat{CR}_1)^{-100,000}, ..., (1+0.00001\ \widehat{CR}_5)^{-100,000})$ would make $\widehat{CA}$ a rather complicated function of the imputation-specific parameter estimates. If, however, we incorporate the imputations in our calculated crash avoidance in the manner exposited in the paragraph following Figure 4-2 (so that the log crash rate is estimated as the average imputation-specific log crash rates, and hence the crash rate as the geometric mean of the imputation-specific crash rates), $\widehat{CA}$ becomes a relatively simple function of the imputation-specific parameter estimates, namely $\widehat{CA}= (1+0.00001\ \exp(X\hat{\beta}))^{-100,000}$ where X gives the model predictors and the entries of $\hat{\beta}$ are the averages of the imputation-

Figure 4-3 depicts the raw crash avoidance estimates for cars and LTVs based on all crashes in 2000-2008. (The leftmost pane depicts the crash avoidance estimates for each type of crash and vehicle, while the rightmost pane depicts the capacity to avoid crashes of any type.) We do not expect the raw estimates to be tremendously accurate, as we lack mileage data on many of the factors that one would intuitively expect to contribute to crash avoidance, such as miles driven drunk, miles driven by drivers of various age or years of driving experience, and by drivers with a history of moving violations. Consequently we will be somewhat circumspect about interpreting our raw (or model estimates) of crash avoidance.

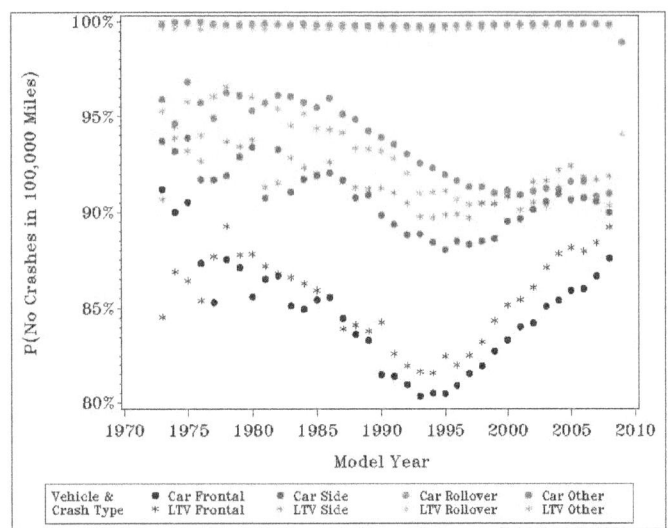 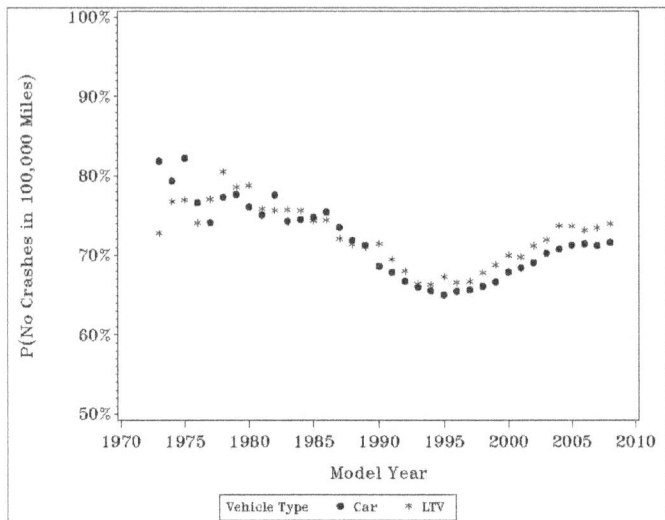

Figure 4-3: Raw Crash Avoidance Estimates

Indeed, the raw estimates of crash avoidance show a curious picture, with seeming declines in crash avoidance prior to model year 1995 and improvements thereafter. However, our analysis will strongly suggest that various factors cloud the relationship between crash avoidance and model year, and that the "true" picture (after disentangling the confounding factors) is a steady improvement in crash avoidance.

4.2 Crashworthiness

Had we only a single imputation, computing raw estimates of crashworthiness would have been straightforward. For instance, to estimate the likelihood that a belted 25- to 65-year-old woman in a rollover of a model year 2000 car with a sober driver sustains at worst a non-incapacitating injury, we would simply compute A/B where B denotes the estimated number of belted 25- to 65-year-old women in model year 2000 car rollovers with sober drivers and A denotes the estimated number among them that sustain at most a non-incapacitating injury. The quantities A and B would be computed as Horvitz-Thompson estimates (i.e., weighted totals) on the dataset formed by combining the crashes in FARS with the non-fatal crashes in GES, using one for the sample weight of each FARS case and using our sole imputation (as well as the known values). However, having multiple imputations forces us to make an order-of-operations choice, and unlike the choice we made for crash avoidance, the manner in which we will incorporate the imputations in crashworthiness estimation is likely to seem unnatural at first glance.

---

specific parameter estimates. The numbers of vehicles in crashes can vary a fair bit among the imputations (e.g., they vary from 3 to 498 (cars) for model year 1976 cars in "other" crashes in 2007), and we have also checked that the alternative calculations would change the crash avoidance estimate by no more than 0.4 percentage points, and by no more than a relative error of 0.5% (calculations omitted).

[32] Consistent with our limitation to production vehicles, we do not estimate crash avoidance for vehicles that are less than -1 year old. NHTSA's FARS and GES databases occasionally contain crashes of vehicles that, based on the coded model year, would appear to be pre-production vehicles (i.e., the difference between their model year and year of crash is at least two). In the 2000-2008 files, there was one such instance, a GES crash of a model year 2006 passenger vehicle in 2004, which we exclude from our study. Even if we wished to study pre-production vehicles, we lack data on miles driven by such vehicles and could not estimate their crash avoidance.

[33] Note that since, e.g., the overall crash rate CR for cars is the sum of the car crash rates CR_j for the four crash types (frontal, rollover, side, and other), the capacity for cars to avoid crashes of any type is equal to $\left(1 + 0.00001 \sum_{j=1}^{4} CR_j\right)^{-100,000}$.

Arguably the most intuitive way to incorporate the imputations would be to average the non-missing values among the imputation-specific estimates (i.e., average the values among $A_1/B_1, \ldots, A_5/B_5$, where A_i and B_i are the analogues to A and B from the previous paragraph in the ith imputation).[34][35] We likely would have computed crashworthiness in this manner had we not developed a statistical model. However, we will develop such a model (in Chapter 5), which will model the log-odds crashworthiness as a linear function of predictors. As a result, it would be inconvenient[36] not to incorporate the imputations in such a manner that the log-odds crashworthiness is equal to the average non-missing imputation-specific log-odds crashworthiness, i.e.

$$\log \frac{CW}{1-CW} = Mean\left(\log \frac{A_1}{B_1 - A_1}, \ldots, \log \frac{A_5}{B_5 - A_5}\right)$$

where $Mean(x_1, \ldots, x_5)$ denotes the mean of the non-missing values among $x_1, \ldots, x_5$.[37] If there is any imputation in which $A_i = 0$ or $A_i = B_i$,[38] the expression on the right hand side, and our calculation of crashworthiness, will not be defined. If, on the other hand, $0 < A_i < B_i$ for each imputation, then the crashworthiness odds will be the geometric mean of the non-missing imputation-specific crashworthiness odds.[39]

$$\frac{CW}{1-CW} = \sqrt[5]{\frac{A_1 \cdots A_5}{(B_1 - A_1) \cdots (B_5 - A_5)}}$$

Applying the inverse odds, we have

$$CW = \left(1 + \sqrt[5]{\frac{(B_1-A_1)\cdots(B_5-A_5)}{A_1\cdots A_5}}\right)^{-1} = \frac{\sqrt[5]{A_1 \cdots A_5}}{\sqrt[5]{A_1 \cdots A_5} + \sqrt[5]{(B_1-A_1)\cdots(B_5-A_5)}}$$

Thus, in order for the crashworthiness log-odds to equal the average of the non-missing imputation-specific crashworthiness log-odds, we are forced to calculate the likelihood that a belted 25- to 65-year-old woman in a rollover of a model year 2000 car with a sober driver sustains at worst a non-incapacitating injury as:[40]

---

[34] That is, B_i denotes the estimated number of belted 25- to 65-year-old women in model year 2000 car rollovers with sober drivers in the ith imputation, and A_i denotes the estimated number among them that sustain at most a non-incapacitating injury. The quotient A_i / B_i will be undefined (missing) when $B_i=0$, e.g. when there are no belted 25- 65-year-old women in a rollover of a model year 2000 car with a sober driver in the ith imputation.

[35] Another natural approach might be to average the numerator and denominators separately, i.e. $(A_1+\ldots+A_5)/(B_1+\ldots+B_5)$. However, this "pooled numerator over pooled denominator" approach cannot be implemented for the model estimates: E.g., the model from the first imputation of the data predicts the log-odds crashworthiness, which only determines the numerator A_1 and denominator B_1 up to scalar multiple.

[36] One example of such inconvenience is illustrated with expressing the model itself. If we estimate crashworthiness as proposed, the crashworthiness model takes the same form as the imputation-specific models and its parameter estimates are the averages of those estimated using each imputation of the crash dataset. That is, the crashworthiness model is logit(CW) = $X\theta$, where $\theta_i = (\beta_{i1}+ \ldots +\beta_{i5})/5$ and β_{ij} is the estimate of the ith parameter from the jth imputation. In contrast, if we estimate crashworthiness by averaging the non-missing values among $A_1/B_1, \ldots, A_5/B_5$, then the crashworthiness model would take a different functional form than the imputation-specific models, namely $CW = Mean(\text{logit}^{-1} X\beta_1, \ldots, \text{logit}^{-1} X\beta_5)$, where β_j is the vector of parameter estimates from the jth imputation.

[37] Note that as advertised, we are estimating the jth imputation-specific crashworthiness as A_j / B_j in this formula.

[38] That is, an imputation (of the crash database) in which among the belted 25- 65-year-old women in model year 2000 car rollovers with sober drivers, either all were incapacitated or killed, or none were incapacitated or killed.

[39] Note that for an imputation-specific crashworthiness odds $CW_i / (1 - CW_i)$ to be defined (non-missing), it is also required that CW_i not equal one, i.e., that there be occupants in the occupant class injured within and beyond the given injury threshold.

[40] It is slightly disappointing that our formulation of crashworthiness requires every imputation to have at least some occupants injured at or below the given KABCO level, *and* at least some injured above the given KABCO level. For instance, suppose that we only imputed twice and that each imputation contained only one member of a rare occupant category (e.g., unbelted 25- to 65-year-old women in rollovers of driverless model year 1970 cars in 2008). In the first imputation, the occupant was incapacitated and in the second, she walked away uninjured. Each imputation produces crashworthiness estimates (e.g., we'd estimate the ability of model year 1970 cars to prevent injury in this occupant class to be 0% in the first imputation, and 100% in the second), so it is somewhat disappointing not to generate a crashworthiness estimate from the pair of imputations. But to us, the inconveniences resulting from defining crashworthiness in a manner other than what we have proposed outweighs this disappointment.

$$CW = \begin{cases} \dfrac{\sqrt[5]{A_1 \cdots A_5}}{\sqrt[5]{A_1 \cdots A_5} + \sqrt[5]{(B_1 - A_1) \cdots (B_5 - A_5)}} & \text{if } 0 < A_i < B_i \text{ for each } i=1,\ldots, 5 \\ \text{undefined} & \text{otherwise} \end{cases}$$

where B_i denotes the estimated number of belted 25- 65-year-old women in model year 2000 car rollovers with sober drivers in the ith imputation, and A_i denotes the estimated number among them that sustain at most a non-incapacitating injury.

Reintroducing the notation from Chapter 3, our general calculation of crashworthiness is as follows: We estimate the ability CW_{cvoz} of the vehicle type v (e.g., model year 2000 car) to protect an occupant class o (e.g., belted 25- to 65-year-old woman with a sober driver) in a crash of type c (e.g., rollovers) against injuries worse than KABCO z as:

$$CW_{cvoz} = \begin{cases} \dfrac{\sqrt[5]{Inj_{cvoz1} \cdots Inj_{cvoz5}}}{\sqrt[5]{Inj_{cvoz1} \cdots Inj_{cvoz5}} + \sqrt[5]{(Occs_{cvo1} - Inj_{cvoz1}) \cdots (Occs_{cvo5} - Inj_{cvoz5})}} & \text{if } 0 < Inj_{cvozi} < Occs_{cvoi} \text{ for each } i=1,\ldots, 5 \\ \text{undefined} & \text{otherwise} \end{cases}$$

where $Occs_{cvoi}$ denotes the Horvitz-Thompson estimate of the number of occupants of type o in vehicles of type v in crashes of type c class in the ith imputation and Inj_{cvozi} denotes the number among them that sustained at most a KABCO z injury.

For another example, the estimated crashworthiness of model year 2000 cars at preventing death in rollovers, supposing that each imputation has fatalities and survivors[41] would be:

$$\dfrac{\sqrt[5]{Surv_1 \cdots Surv_5}}{\sqrt[5]{Surv_1 \cdots Surv_5} + \sqrt[5]{Killed_1 \cdots Killed_5}}$$

where $Surv_i$ and $Killed_i$ are the numbers of survivors and fatalities in among occupants in rollovers of model year 2000 cars in the ith imputation.

What about our crash-type-independent version of crashworthiness (e.g. the ability to protect occupants in *any* type of crash)? We simply estimate the generalized notion in the same manner as we did the specific. That is, we estimate the ability CW_{voz} of the vehicle type v to protect an occupant class o in a crash of *any* type against injuries worse than KABCO z as:

$$CW_{voz} = \begin{cases} \dfrac{\sqrt[5]{Inj_{voz1} \cdots Inj_{voz5}}}{\sqrt[5]{Inj_{voz1} \cdots Inj_{voz5}} + \sqrt[5]{(Occs_{vo1} - Inj_{voz1}) \cdots (Occs_{vo5} - Inj_{voz5})}} & \text{if } 0 < Inj_{vozi} < Occs_{voi} \text{ for each } i=1,\ldots, 5 \\ \text{undefined} & \text{otherwise} \end{cases}$$

where $Occs_{voi} = \sum_{k=1}^{5} Occs_{cvoki}$ and $Inj_{voi} = \sum_{k=1}^{5} Inj_{cvoki}$.[42]

The raw estimates indicate steady improvements in crashworthiness as a function of model year. In Figure 4-4, which presents crashworthiness estimates for cars and LTVs based on crash data from the calendar years 2000-2008, we recall that the KABCO scale is: O = no injury, C = possible injury, B = non-incapacitating injury, A = incapacitating injury. Thus the green dots and stars in Figure 4-4 give the likelihood of escaping a crash uninjured, while the purple symbols plot the chance of experiencing at most a non-incapacitating injury, and the red give the chance of surviving a crash (which is quite high). The blue symbols are a bit more

---

[41] While we do not impute any unknown levels of injury as death, elements such as model year, crash type, and vehicle type are imputed, and so the numbers of fatalities (and survivors) in the occupant class can vary among imputations.

[42] In Chapter 3 we defined our crash-type-independent crashworthiness to be $\sum_{k=1}^{5} p_k CW_k$, where k indexes over our five crash types (frontal, near side, far side, rollover, and other), $(p_1, \ldots p_5)$ is a distribution of these crash types, and $CW_1, \ldots CW_5$ are the crash-type-specific versions of crashworthiness. This born of necessity, in that we needed to specify a mechanism for generating crashes, or at least the crash type distribution in such a mechanism, in the underlying probabilistic experiment. However, we can estimate crash-type-independent crashworthiness directly from the crash data in the same manner we used for the crash-type-specific versions, as we have detailed in the formula for CW_{voz}, which reflects the distribution of crash types in the crash database.

amorphous to interpret as they give the likelihood of escaping with only a "possible injury". This KABCO code *C* reflects a collection of often State-specific codes on the police accident report that indicate a degree of uncertainty (at the time the police officer is filling out the report) about the severity of a person's injury.[43][44] (National Highway Traffic Safety Administration, 2008)

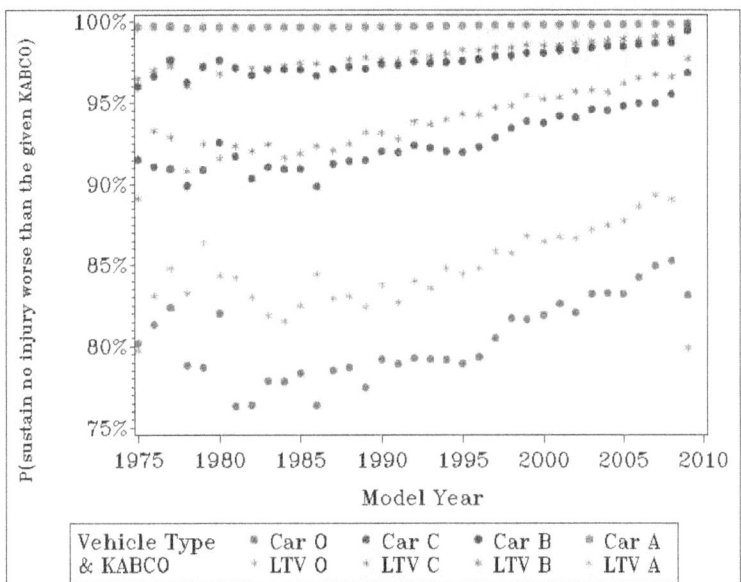

Figure 4-4: Raw Crashworthiness Estimates

However, the estimates in Figure 4-4 are only controlled for vehicle type, so the apparent improvement could be confounded by factors such as crash year, belt use, and crash type. For instance, SUV rollover crashes increased steadily during 2000-2008,[45] and so some of the crashworthiness estimates in Figure 4-4 might be understated, since rollovers tend to be more severe than other crash types.[46] Conversely, belt use increased each year during the 2000-2008 time period, and so the crashworthiness estimates for newer vehicles might be overstated. A major aim of this report is to isolate the impact of model year on crashworthiness and crash avoidance from the impacts of other factors such as belt use and crash type, an aim we intend to address through our statistical models.[47]

---

[43] States use different reporting codes to characterize injury severity on accident reports. A number of States have a code specifically for "possible injury", while others have a code for "complaint of pain", "probable-not apparent [injury]", or other characterizations. See (National Highway Traffic Safety Administration, 2008) for the reporting codes in each State that are coded in FARS and GES as KABCO *C*.

[44] Note that the different criteria used by States to determine which crashes are to be reported to the police affects our crashworthiness estimates. For instance, a State that uses a property damage threshold of $1,000 likely possesses fewer KABCO *O* cases in GES (and FARS) than a State that uses a $500 threshold. Unfortunately we do not have information with which to adjust for such differences, nor assess their impact.

[45] The number of SUV rollovers reported to the police increased from 67,000 in 2000 to 83,000 in 2008. (NHTSA's National Center for Statistics and Analysis, 2001), (NHTSA's National Center for Statistics and Analysis, 2009) For the numbers of rollovers in each intervening year, see the corresponding Traffic Safety Facts publications for these data years.

[46] For instance, model year 2008 LTVs only appear in the 2008 crash year, when LTV rollover incidence was relatively high, while model year 2000 LTVs appear in crash years 2000-2008, when fewer LTVs rolled over on average. If, hypothetically, the ability of a model year 2008 LTV to protect against, e.g., non-incapacitating injuries was *identical* to that of model year 2000 LTVs and the sole manner in which the crashes experienced by LTVs changed during the 2000-2008 calendar years was the increase in rollovers, then our estimates would likely indicate the model year 2008 LTVs to provide diminished protection (against these types of injuries), compared to the model year 2000 LTVs, as the types of crashes against which the model year 2008 LTVs are assessed are more severe (and so, more difficult to perform well against) than those used to assess the model year 2000 LTVs. In short, the estimates in Figure 4-4 could be clouded by increases (or decreases) in the severity of the typical crash during the 2000-2008 calendar year time frame, perhaps as a result of shifts in the types of crashes that occur.

[47] While crashworthiness per se should be independent of driver behavior, the crashworthiness estimates in the references charts are derived from crash data and so are influenced by the types of crashes motorists get into. If, for example, through improved driving behavior, motorists get into less severe crashes, the crash-based crashworthiness estimates will decline, even if vehicles have become no more crashworthy.

Appendix 8.1 contains plots of the raw crashworthiness estimates for a selection of the following crash, vehicle, and occupant types, based on crash data pooled from calendar years 2000-2008. [48][49]

Table 4-1: Values of Categorical Variables

Variable	Values
Crash type	Rollover crashes, frontal impacts, near side impacts, and far side impacts[50]
Vehicle type	Passenger cars, LTVs[51]
Driver alcohol	Sober driver, non-sober driver, no driver
Restraint use	Restrained, unrestrained[52]
Occupant age category	< 14 years, 14-24 years, 25-65 years, > 65 years
Occupant gender	Female, male

---

[48] By definition, the crashworthiness for KABCO level K is 1, i.e. at worst one dies, and so KABCO K does not appear in our crashworthiness plots. Note that this does *not* mean that fatalities are excluded from the crashworthiness estimates: Killed occupants, and occupants injured at every severity level, contribute to *every* crashworthiness estimate, by contributing to the assessed probability of being injured at no worse than a given injury threshold.

[49] Sometimes police reports are more sparsely filled out for uninjured occupants than for injured occupants. For instance, the space for occupant age might be left blank with higher frequency for occupants who are not injured. Consequently our estimates involving uninjured occupants could involve a greater contribution from imputed values than do other estimates.

[50] We determine the point of impact in non-rollover crashes from the General Area of Damage variable for GES cases and the Principal Impact Point variable for FARS. Near side and far side impacts are characterized from the perspective of the occupant. For instance an occupant on the driver's side of the vehicle in a driver's side impact is in a near side crash. Occupants in the center seating position were considered to be in far side crashes.

[51] LTVs (light trucks and vans) comprise sport utility vehicles, vans, and pickup trucks. We use the passenger car and LTV classifications as provided in (Chang, 2009) and (National Highway Traffic Safety Administration, 2008).

[52] We categorize restraint use as restraint used or not used. We do not attempt to incorporate improper use of restraints. Rather, we categorize occupants as having used restraints if they are coded as having used restraints properly or improperly (or were imputed to have used restraints based on our imputation algorithm).

5. Model Fitting

5.1 The Crash Avoidance Model

Outlier Analysis and Model Training Data

We see in the following figure that brand new vehicles (those with age -1) are outliers, and will exclude them from the model fitting.[53]

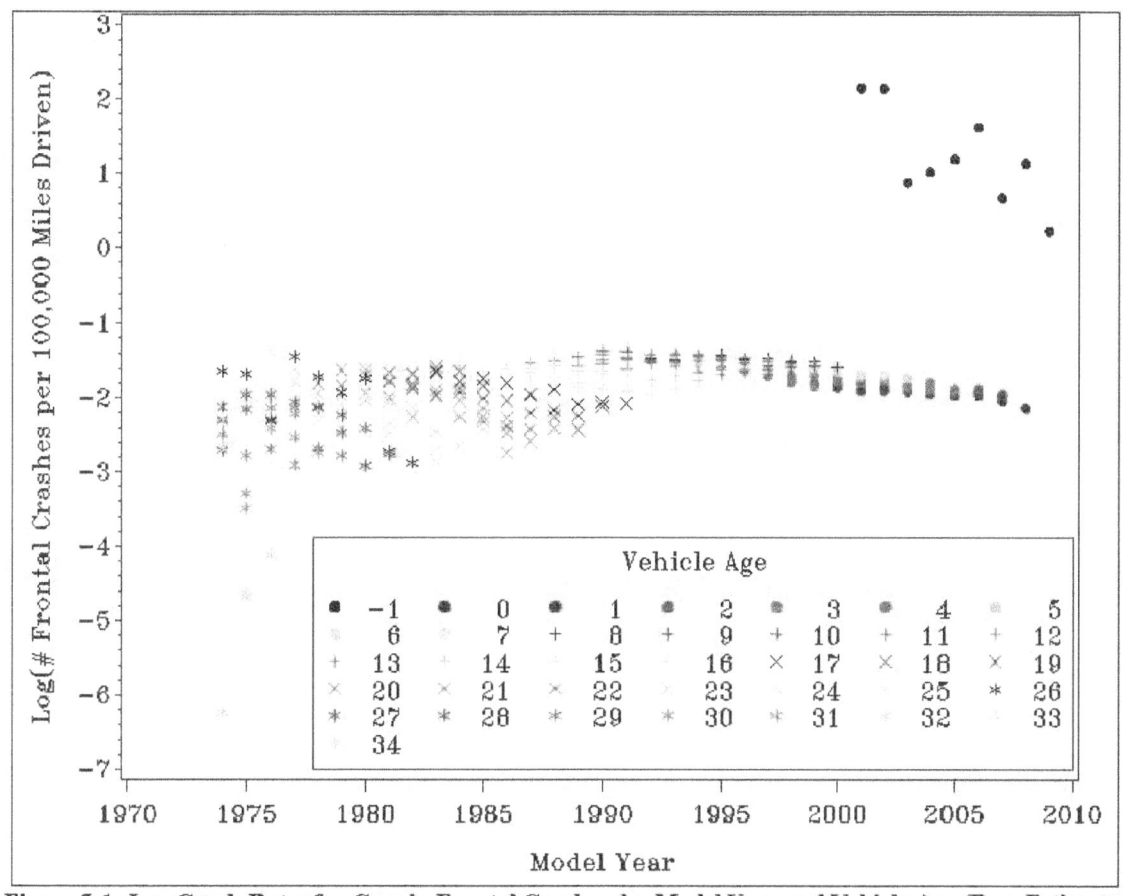

Figure 5-1: Log Crash Rates for Cars in Frontal Crashes, by Model Year and Vehicle Age (Raw Estimates)

We note in Figure 5-1 that the data become more scattered as vehicle age increases, which makes sense because we have less data on, say, 30-year-old vehicles than we do on 10-year-old ones. Thus, we wish to implement some vehicle age limit and somewhat arbitrarily choose a cutoff of 20 years for vehicle age. Similar exploratory data analysis uncovered no additional outliers. All together, our model training data consists of estimated crashes and miles driven by vehicles ages 0 through 20 during the calendar years 2000-2008.

Preliminary Effects Involving Vehicle Age and Crash Year

The previous chart also suggests that crash avoidance is affected by vehicle age (or crash year), and so should be included as a factor in the model. The plot below with vehicle age on the x-axis suggests that crash avoidance is a quadratic function of vehicle age and a linear function of crash year, with no apparent interaction between vehicle age and crash year.[54] This plot also suggests the

---

[53] While the figure depicted only concerns cars in frontal crashes, similar patterns are seen in other crash and vehicle types.
[54] Likewise, the figure depicted concerns cars in frontal crashes, but similar patterns are seen in other crash and vehicle types. The curves in the figure are least squares quadratic fits. The linear crash year relationship is suggested by the near-uniform vertical displacement of the crash year curves. (Although it appears in Figure 5-2 that the effect of some years may be stronger than others, with a possibly stronger change in 2005-2006 than in other years, the Shapiro-Wilk test finds the estimates of the calendar year effect from the quadratic fits are consistent with a random sample from a normal distribution. That is, the calendar year effect depicted in Figure 5-2 is indistinguishable from a constant decrease in the log crash rate, plus random noise.)

independently interesting finding that as a vehicle ages, the driver cohort shifts to more, and then less, crash-prone drivers. The plot with model year on the *x*-axis shows that it will be easier to uncover the form of the crash avoidance model using vehicle age as a predictor than using model year as a predictor.[55]

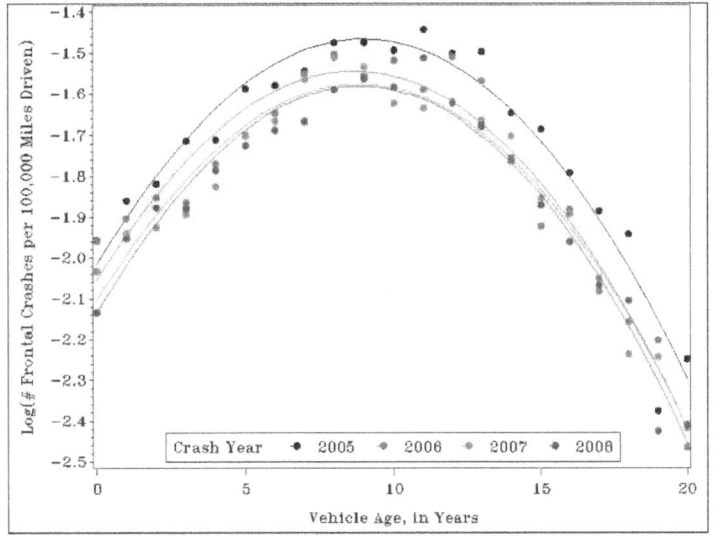 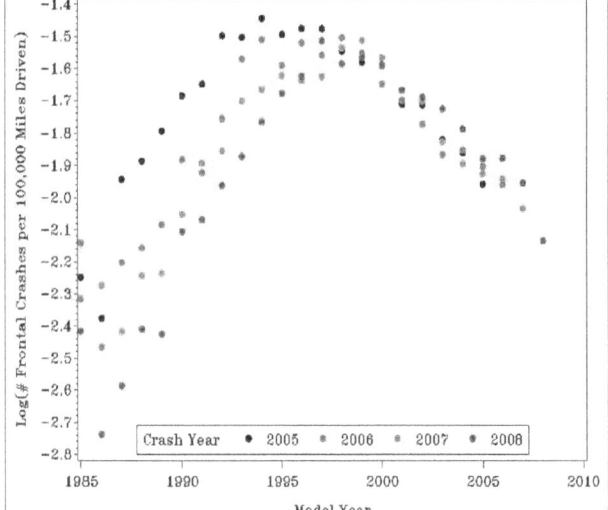

Figure 5-2: Log Crash Rates for Cars in Frontal Crashes, by Vehicle Age and Crash Year (Raw Estimates With Quadratic Fits)

Figure 5-3: Log Crash Rates for Cars in Frontal Crashes, by Model Year and Crash Year (Raw Estimates)

Thus, we shall include linear and quadratic vehicle age terms and a linear crash year term as preliminary effects in the crash avoidance model.

Preliminary Effects Involving Vehicle Type and Crash Type

We note in the next figure that fixing the year of the crash, the flatness of the vehicle age effect appears to be affected by the crash type and vehicle type. This suggests that holding crash year constant, the linear and quadratic vehicle age terms depend on crash type, vehicle type, and their interaction. That is, using the shorthand CY, CT, VT, and VA for the crash year, crash type, vehicle type, and vehicle age factors respectively, we should include the following terms as preliminary effects: CT, VT, VA, CT*VT, CT*VA, VT*VA, CT*VT*VA, VA^2, $CT*VA^2$, $VT*VA^2$, $CT*VT*VA^2$.

---

[55] Although the relationship to model year also appears somewhat quadratic to the naked eye, we note that SAS chooses a linear function as the least-squares quadratic fit and informs us that the quadratic solution is not unique.

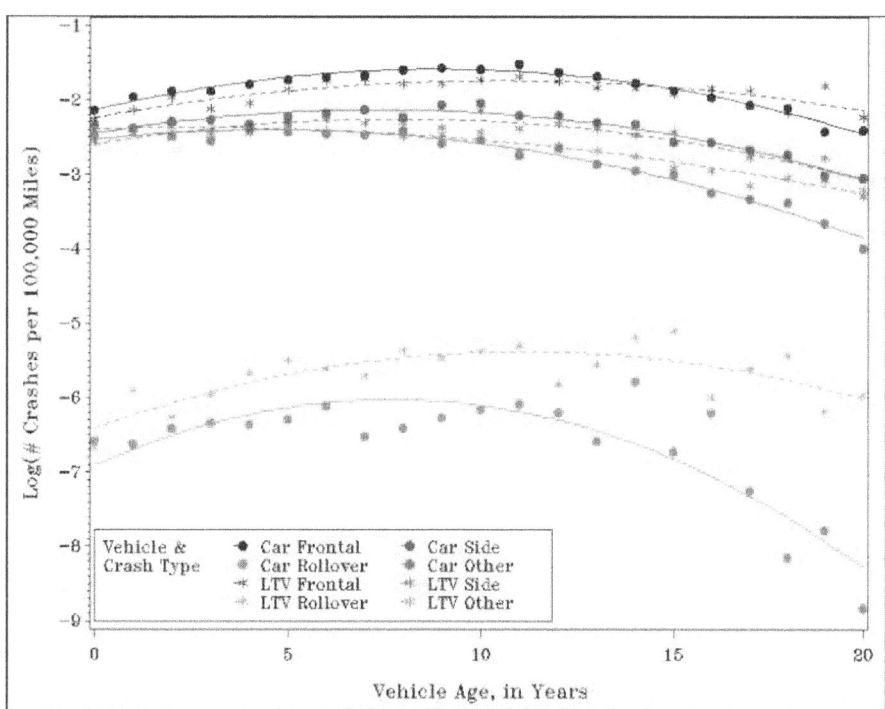

Figure 5-4: Log Crash Rates in 2008, by Vehicle Type and Crash Type (Raw Estimates With Quadratic Fits)

The next plot suggests that at each vehicle age, the crash type and vehicle type impact the slope and intercept of the crash year effect. Thus, the model should also include the terms: CT*CY, VT*CY, CT*VT*CY.[56]

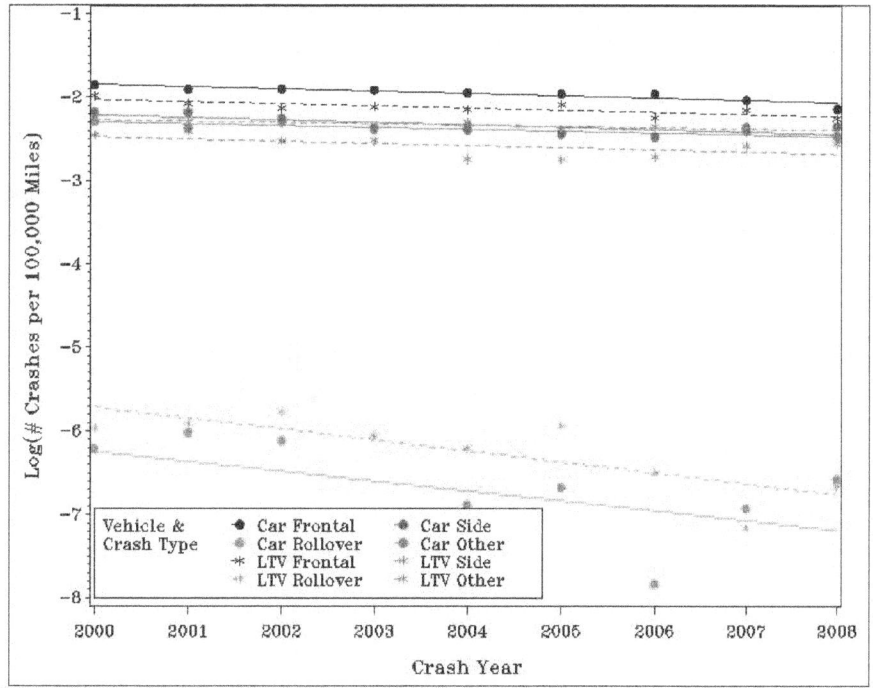

Figure 5-5: Log Crash Rates for Vehicles at Age 0, by Vehicle Type and Crash Type (Raw Estimates With Linear Fits)

All together, we have a preliminary model form of:

---

[56] The fitted curves (lines) are linear in this plot since we are plotting against crash year.

$$\log(\text{crash rate}) \sim \text{CY, CT, VT, VA, CT*VT, CT*CY, VT*CY, CT*VT*CY, CT*VA, VT*VA, CT*VT*VA, VA}^2, \text{CT*VA}^2, \text{VT*VA}^2, \text{CT*VT*VA}^2$$

A Geometric Perspective

We pause to convey what the crash avoidance surface looks like. For a given crash type and vehicle type, we are modeling the log crash rate as linear in calendar year and quadratic in vehicle age, which, completing the square, we could write as:

$$\log(\text{crash rate}) = a(x-b)^2 + c(y-2000) + d$$

where x and y denote vehicle age and calendar year, respectively, and the constants a, b, c, and d depend on the type of crash and vehicle under consideration (e.g., frontal crashes of cars). This is a parabolic cylinder, distorted to decrease with increasing calendar year. Figure 5-6 provides a picture of the general shape these crash- and vehicle-dependent surfaces.

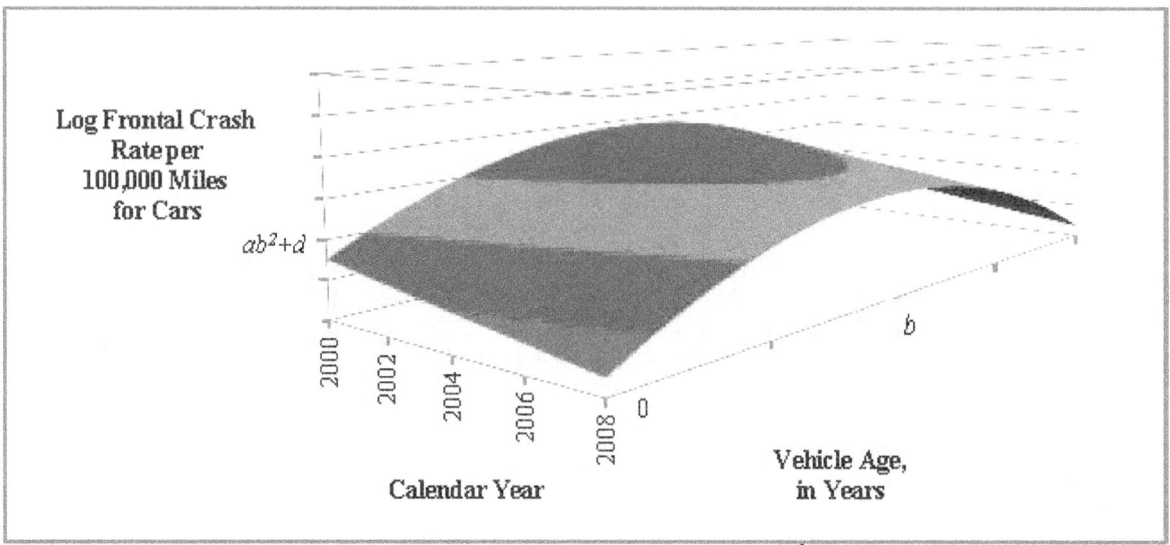

Figure 5-6: The Shape of the Log Crash Rate Surface $a(\textit{Vehicle Age} - b)^2 + c(\textit{Calendar Year} - 2000) + d$

We are fitting a model based on estimated crashes and miles traveled for vehicles that were ages 0 through 20 years in calendar years 2000-2008, i.e., over a rectangular region in the xy-plane.

The crash rate surface would form a non-parabolic stretching of the log crash rate surface, as this model (again, for a given crash and vehicle type) is: crash rate = $\exp(a(x-b)^2 + c(y-2000) + d)$. The crash rate surface would resemble Figure 5-7.

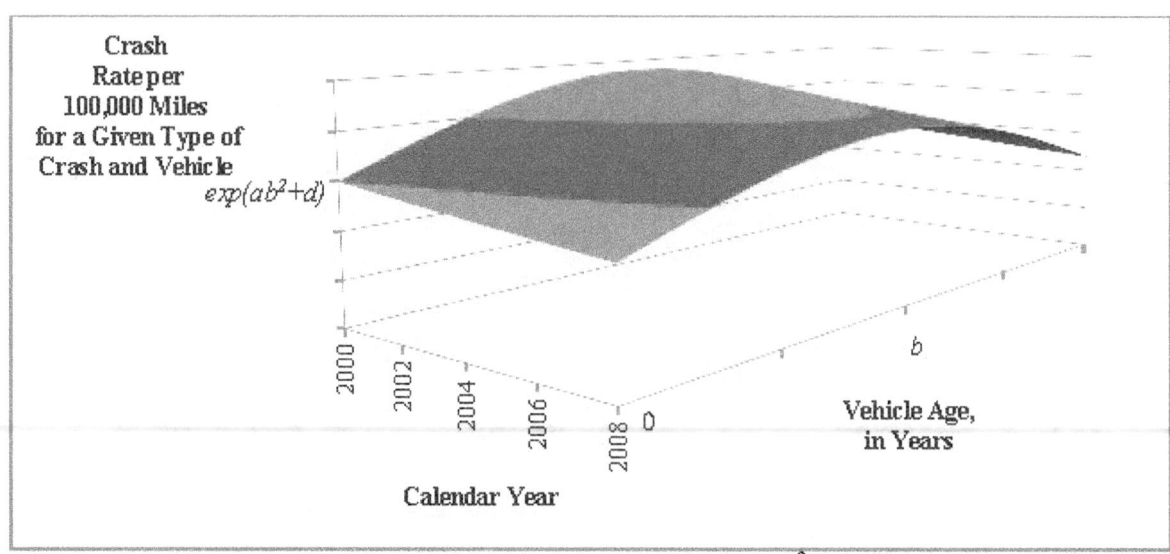

Figure 5-7: The Shape of the Crash Rate Surface $\exp(a(\textit{Vehicle Age} - b)^2 + c(\textit{Calendar Year} - 2000) + d)$

The crash avoidance surface (for the same crash and vehicle type) would be defined by:

$$\text{crash avoidance} = (1 + 0.00001\ exp(a(x-b)^2 + c(y-2000) + d))^{-100,000}$$

and would resemble Figure 5-8.

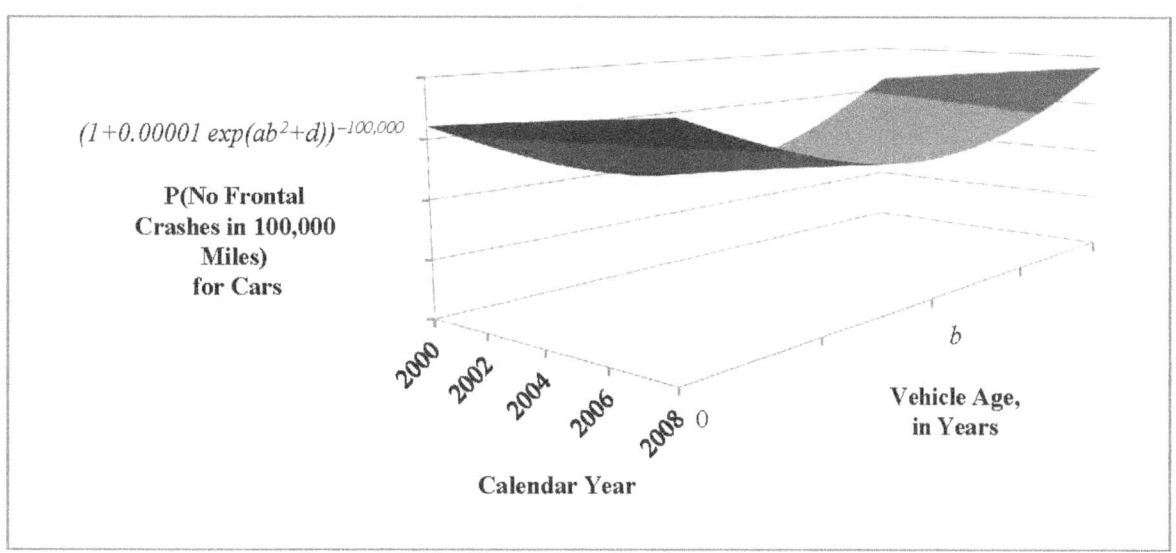

Figure 5-8: The Shape of the Crash Avoidance Surface
$(1 + 0.00001\ exp(a(Vehicle\ Age - b)^2 + c(Calendar\ Year - 2000) + d))^{-100,000}$

Recalling that overall crash rate CR for cars or LTVs[57] is the sum of the car (respectively, LTV) crash rates CR_j for the four crash types (frontal, rollover, side, and other), the crash rate surface reflecting the rate of car involvements in crashes of any type (as a function of year of driving and vehicle age) would be a non-linear combination of the surfaces like that in Figure 5-7, each with particular values of a, b, c, and d.[58] Likewise, the overall crash avoidance surface would be a non-linear combination of crash type-specific surfaces like that in Figure 5-8.[59] Chapter 6 will depict these (model-estimated) surfaces.

Refinement of Model Factors

We return to building the crash avoidance model. The effects CY*VT, CY*VT*CT, VA*VT, VA*VT*CT, VA^2*VT, VA^2*VT*CT are not significant, so are dropped for the final model.[60]

---

[57] That is, the rate of car involvements in (any type of) crash to the collective miles driven.
[58] Namely, the equation for the overall crash rate surface would be $\sum_i exp(a_i (x-b_i)^2 + c_i (y-2000) + d_i)$.
[59] Likewise, the equation for the overall crash avoidance surface would be $(1 + 0.00001 \sum_i exp(a_i (x-b_i)^2 + c_i (y-2000) + d_i))^{-100,000}$
[60] We used the relative crash year (i.e., crash year minus 2000) for the crash year effect to aid in model convergence. The model does not converge using crash year, a fact that we think is due to SAS's initial guess in applying Newton's method to maximize the likelihood function.

Table 5-1: Type 3 Results for Preliminary Effects, Using a Negative Binomial Model

Effect	Average p-Value Among the Type III Tests For the Five Imputations
CT	<0.0001
CY	<0.0001
VA	<0.0001
VA^2	<0.0001
VT	0.0251
CY*CT	<0.0001
CY*VT	0.2433
CY*VT*CT	0.2857
VA*CT	<0.0001
VA*VT	0.6345
VA*VT*CT	0.8763
VA^2*CT	0.0002
VA^2*VT	0.1927
VA^2*VT*CT	0.8534
VT*CT	<0.0001

The Choice of the Negative Binomial Model

The dispersion parameter is significantly different from (and larger than) zero, confirming our choice to model crashes via a negative binomial distribution, instead of Poisson.[61]

Table 5-2: Dispersion Parameter for Preliminary Effects

Parameter	Estimate	Standard Error	Wald Confidence Interval	
			Lower Bound	Upper Bound
Dispersion	0.0508	0.0019	0.0470	0.0545

The Final Model

All final effects have significant Type III results, and the dispersion parameter is still needed (Tables 5-3 and 5-4), so this is our final model.

The Final Crash Avoidance Model (5-1)

$$\log(\text{crash rate}) \sim CY, CT, VA, VA^2, VT, CY*CT, VA*CT, VA^2*CT, VT*CT$$

That is, the terms in our model consist of Crash Type, Calendar Year, Vehicle Age, Vehicle Age^2, Vehicle Type, and the interaction of the last four of these with Crash Type.

---

[61] The dispersion estimate presented in Table 5-2 is the maximum likelihood estimate of the quantity $(\sigma^2-\mu)/\mu^2$, where μ and σ^2 are the mean and variance of the random variable Y defined by the number of crashes in 100,000 miles. That is, $\text{Var}(Y) = \mu + \kappa\mu^2$, where κ is the dispersion parameter, and overdispersion is indicated by positive estimate for κ. The figures in this table are averages from the five imputation models.

Table 5-3: Type 3 Results for the Final Crash Avoidance Model

Effect	Average p-Value Among the Type III Tests For the Five Imputations
CY	< 0.0001
CT	< 0.0001
VA	< 0.0001
VA^2	< 0.0001
VT	0.0059
CY*CT	< 0.0001
VA*CT	< 0.0001
VA^2*CT	< 0.0001
VT*CT	< 0.0001

Table 5-4: Dispersion Parameter for the Final Crash Avoidance Model

Parameter	Estimate	Standard Error	Wald Confidence Interval Lower Bound	Wald Confidence Interval Upper Bound
Dispersion	0.0511	0.0019	0.0473	0.0549

While the model fits the data well in some circumstances, such as that depicted in Figure 5-9, Figure 5-10 indicates the model could stand improvement.

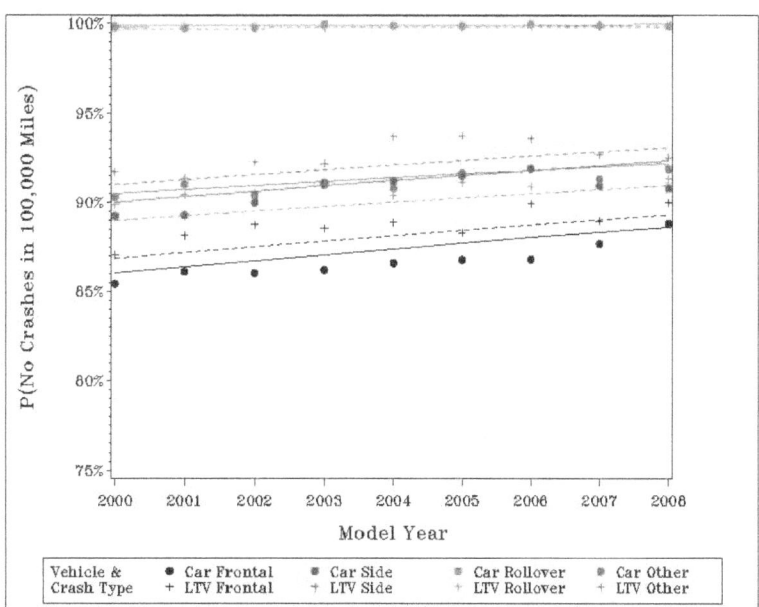

Figure 5-9: Crash Avoidance for Age Zero Vehicles, by Model Year, Vehicle Type, and Crash Type (Raw and Model Estimates)

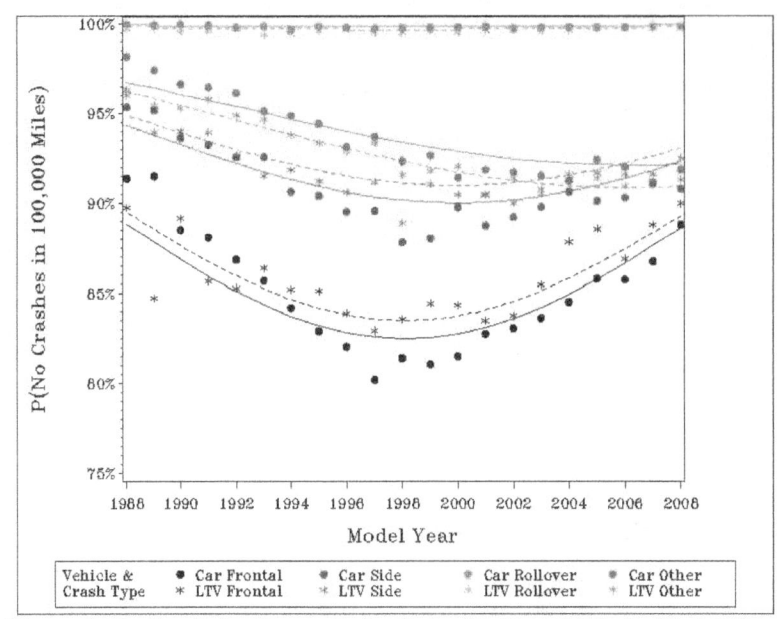

Figure 5-10: Crash Avoidance in the Calendar Year 2008, by Model Year, Crash Type, and Vehicle Type (Raw and Model Estimates)

The Model Parameters

Having chosen our crash avoidance model, the following chart depicts the parameter estimates, expressed as multiplicative effects on the reference group's crash rate.[62] Our reference group comprises age zero cars in frontal crashes in 2000, which has a crash rate of 0.15 (frontal) crashes per 100,000 miles traveled. We note that the parameter estimate for rollovers in LTVs is particularly large, at twice the crash rate of the reference group. The numerical values of the model parameter estimates appear in Table A-2 of the Appendix.

---

[62] Figure 5-11 depicts the geometric mean of the imputation-specific estimated multiplicative effects, i.e., the exponentiated average of the imputation-specific effects on the log crash rate. For example, the blue dot for Crash Year in Figure 5-11 is equal to $exp(\sum_i \beta_i/5)$ where the β_i are the five parameter estimates for the coefficient of Crash Year in the model of the log crash rate. When we compute predicted values of crash avoidance, we average the imputation-specific predictions, which is equivalent to applying the model of the form (5-1) whose parameter estimates are the averages of the corresponding imputation-specific parameter estimates.

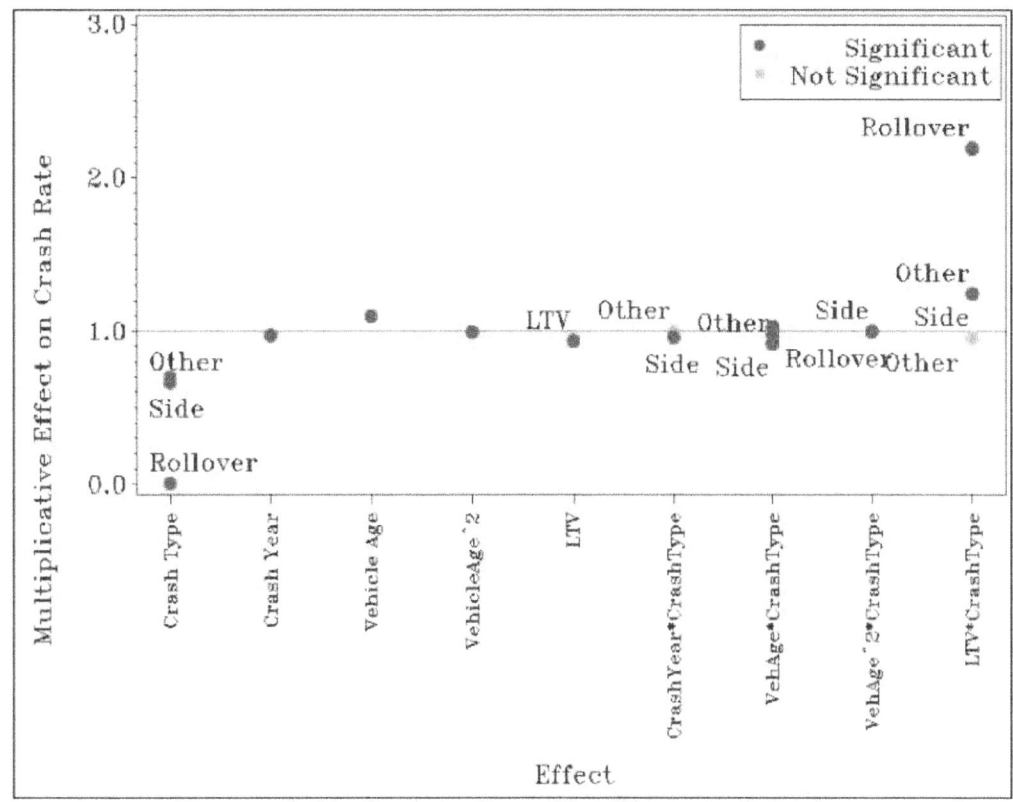

Figure 5-11: The Crash Avoidance Model Parameter Estimates

We note also that consistent with the downward parabola in Figure 5-2, the multiplicative effect of VA^2 on the crash rate is less than one (but just slightly, at 0.995), i.e., the additive effect on the log crash rate is negative.

Model Fit

Although the residuals show some unexplained variation, the plots of raw and model estimates indicate a good fit.

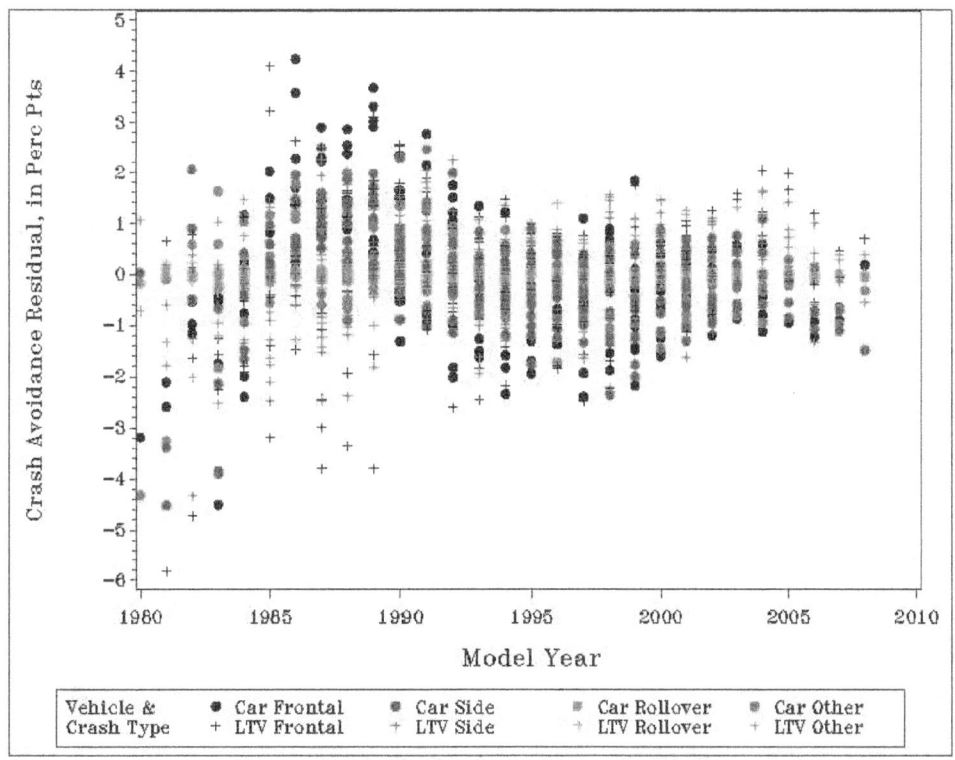

Figure 5-12: The Crash Avoidance Residuals

Most of the parameter estimates have a relative error less than 20 percent, with less than 15 percent of the variance occurring between imputations. The parameters with more than 20 percent relative error are small (absolute value of the additive effect on the log crash rate being less than 0.1), when a larger relative error is understandable. Thus, our parameter estimates generally have low variability and perturbing the data via imputation yields generally similar parameter estimates, so there do not appear to be relationships among the effects.

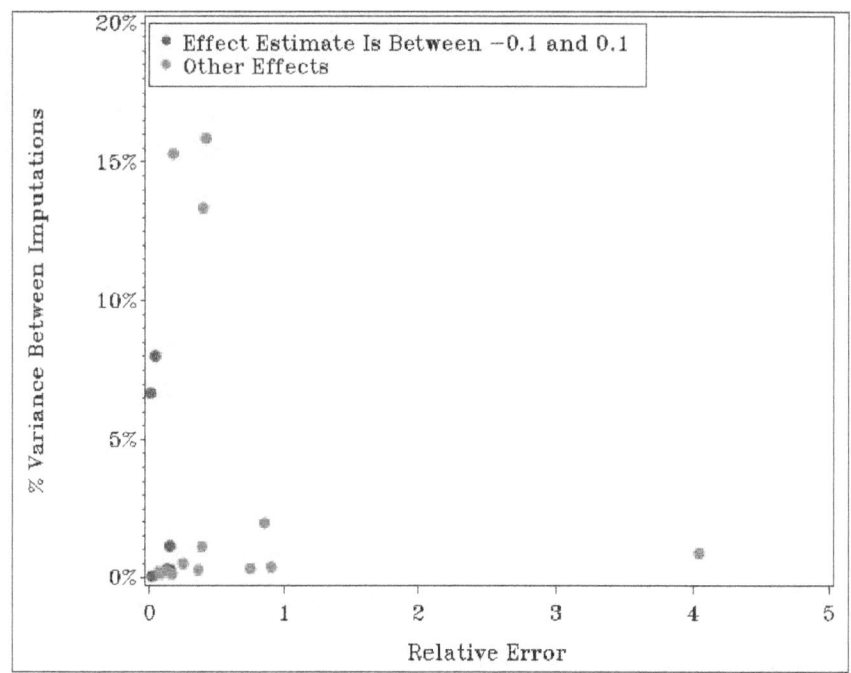

Figure 5-13: Crash Avoidance Model Variances

Adequacy of Model Controls and Possible Future Model Refinements
Our crash avoidance model attempts to control for driver effects by including the Vehicle Age variable. Consequently, our model filters out the human effect on crash avoidance to the extent that drivers of the same age cars (respectively, light trucks) drive similarly (in terms of factors such as degree of recklessness, drunk driving, speeding, and driving experience). Unfortunately we do not have data on, e.g. miles driven drunk or miles logged by experienced drivers, which would help us to better model the human contribution to crash propensity.

The model has no environmental controls.[63] Although it is quite possible that factors such as road conditions, visibility, and road design play a role in crash propensity, we lack data on miles driven under various types of road conditions, visibility, and road design. It is conceivable to us that the factors in our model might be adequate, and we cannot think of other factors (on which we have data) that would improve the isolation of factors related to vehicles.

The model uses broad vehicle categories, and a possible future improvement would be to use finer vehicle categories (such as small cars, etc), to the extent it is possible to obtain mileage for the categories from the NHTS data. We would also be interested to use mileage at the time of the crash, which is available in NHTSA's Crashworthiness Data System, to enhance the information supplied by the NHTS. Other possible refinements would be to include occupant travel characteristics from the NHTS (such as driver age) and incorporate sampling error from the NHTS into the crashworthiness estimates and model.

---

[63] Although the model contains the variable Calendar Year, we shall assess improvements to crash avoidance by advancing the Calendar Year variable, and so our assessment will have no environmental controls.

5.2 The Crashworthiness Model

Rejection of Effects Involving Vehicle Age and Crash Year

Figure 5-14 indicates that crashworthiness is independent of vehicle age (or crash year). The plot shows no systematic shift from light to dark dots, or vice versa, in a given model year.[64] Thus, we will exclude vehicle age and crash year as factors in our crashworthiness model.[65]

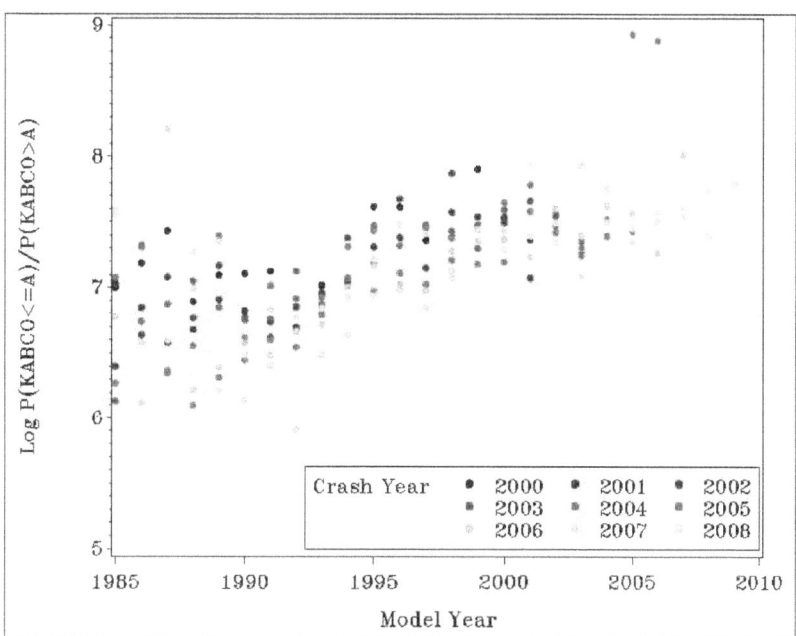

Figure 5-14: Log Odds Crashworthiness of KABCO A Injury for Belted 25- to 65-Year-Old Women in Frontal Car Crashes With Sober Drivers (Raw Estimates)

This leaves us with one ordinal predictor (KABCO), and one interval predictor (model year), and six non-ordinal categorical predictors, as indicated in Table 5-5.

---

[64] Statistical tests also support this assertion: The Shapiro-Wilk test finds the per-year effects (i.e., the differences in the log odds survival in successive calendar years, as estimated from the data in Figure 5-14) to be consistent with a random sample from a normal distribution. The ANOVA F-test (of the model predicting the per-year effect from calendar year, treated as a class variable) finds calendar year to be uninformative for predicting the per-year effect. That is, the calendar years don't seem to have discernible effects. The ANOVA F-test of the intercept-only model (predicting a constant per-year effect) finds the calendar year effects in Figure 5-14 to be consistent with random noise. Although Figure 5-14 depicts the special case (of A-injured belted 25- to 65-year-old women in frontal car crashes with sober drivers), similar results are seen in other slices of the data.

[65] Figure 5-14 plots the log crashworthiness. All logarithms in this paper are natural logarithms (i.e., have base e).

Table 5-5: Crashworthiness Model Predictors

Variable	Values
Crash type	Frontal, near side, far side, rollover, and other
Vehicle type	Car, LTV
Driver alcohol	Alcohol involved, alcohol not involved, no driver
Restraint use	Restrained, unrestrained
Occupant age category[66]	< 14 years, 14-24 years, 25-65 years, > 65 years
Occupant gender	Female, male
KABCO[67]	O, C, B, A
Vehicle model year[68]	1985, 1986, ..., 2008

Model Form

Examining the plots of the log odds of crashworthiness, log odds of injury, and log odds of survival[69] versus model year, for the various crash types, vehicle types, driver alcohol status, and occupant restraint use, age group, and gender, reveals that the relationship between log odds crashworthiness and model year is at least as linear (for a given KABCO) as that between the log odds of injury and model year or log odds survival and model year. Figures 5-15 and 5-16 depict one set of plots in which the differences in linearity are particularly striking. The lines in these charts are least-squares fits. Consequently, we reject the generalized logistic model.[70]

---

[66] We categorize occupant ages, rather than using age as an ordinal or continuous predictor, because our data is too sparse to estimate crashworthiness reliably for individual ages. Our choice of age ranges to use in the categories is somewhat arbitrary, although it might be a fairly common choice among traffic studies. We do not mean to imply a priori by our choices that crashworthiness is necessarily statistically significantly different for occupants under 14 years of age, compared to 14- to 24-year-olds. However, our results will find statistically significant differences in crashworthiness in various age categories.

[67] We omit KABCO K, as the crashworthiness at this level is always 100 percent (i.e., the worst outcome is death). As noted earlier, occupant fatalities still contribute to the crashworthiness estimates for KABCO levels $O, C, B,$ and A.

[68] We use individual model years, rather than using categories such as model years 1985-1989, 1990-1994, etc., with the hope that we can estimate a per-year effect in our crashworthiness model (which will turn out to be the case).

[69] The log odds of injury at injury level z is defined to be $log(P(Injury=z)/P(no\ injury))$, while the log odds of survival is defined to be $log(P(Injury=z)/P(fatality))$. In estimating the log odds injury and log odds survival, we treat our five hotdeck imputations in a manner consistent with the approach we took for crashworthiness. Namely, for a given crash type, vehicle types, driver alcohol status, and occupant restraint use, age group, and gender, we estimate $P(Injury=z)$ to be the average of the non-missing values among $A_1/B_1, ..., A_5/B_5$, where B_i denotes the number of occupants of the given crash type, vehicle types, driver alcohol status, and occupant restraint use, age group, and gender in the ith imputation of our crash database and A_i denotes the number of such injured at level z. We then estimate the log injury odds and log survival odds to be $log(P(Injury=z)/P(Injury=z_0))$, where z_0 is "no injury" or "fatality", respectively.

[70] A generalized logistic model would model the log-odds of a given injury level compared to a reference injury level, as a function of model year and our categorical variables (crash type, vehicle type, etc). That is, the model's dependent variable would have the form $log(P(Injury=z)/P(Injury=z_0))$, where z ranges among the injury levels and z_0 denotes the reference level). The most natural reference levels would be z_0= no injury (in which case, we are modeling the log odds injury) or z_0= fatality (producing the log odds survival). Figures 5-16 and 5-17 demonstrate that neither of these is very linear in model year. Thus, we could either model the log-odds crashworthiness (which as depicted in Figure 5-15, looks linear in model year) or attempt to decipher a non-linear relationship for the log injury odds or log survival odds, and we reject the latter as less practical.

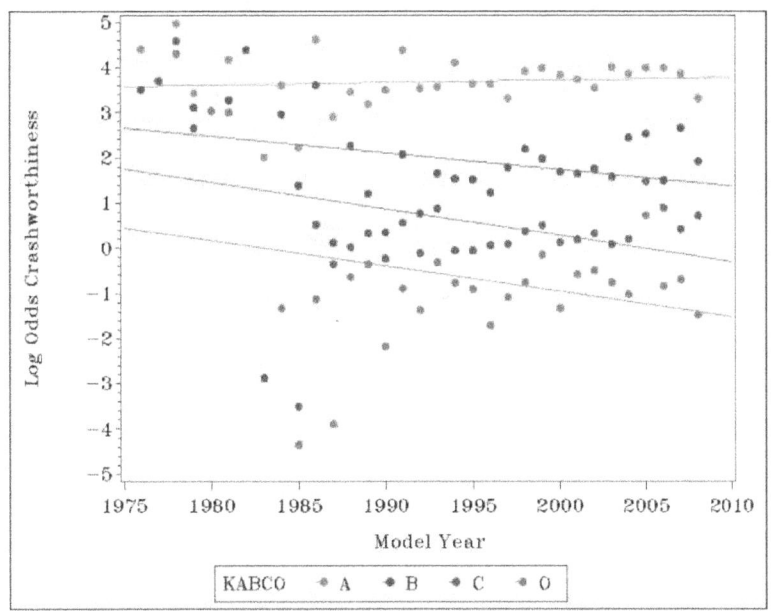

Figure 5-15: Log Odds Crashworthiness for Belted 25- to 65-Year-Old Women in LTV Rollovers With Sober Drivers in 2000-2008 (Raw Estimates With Least-Squares Linear Fits)

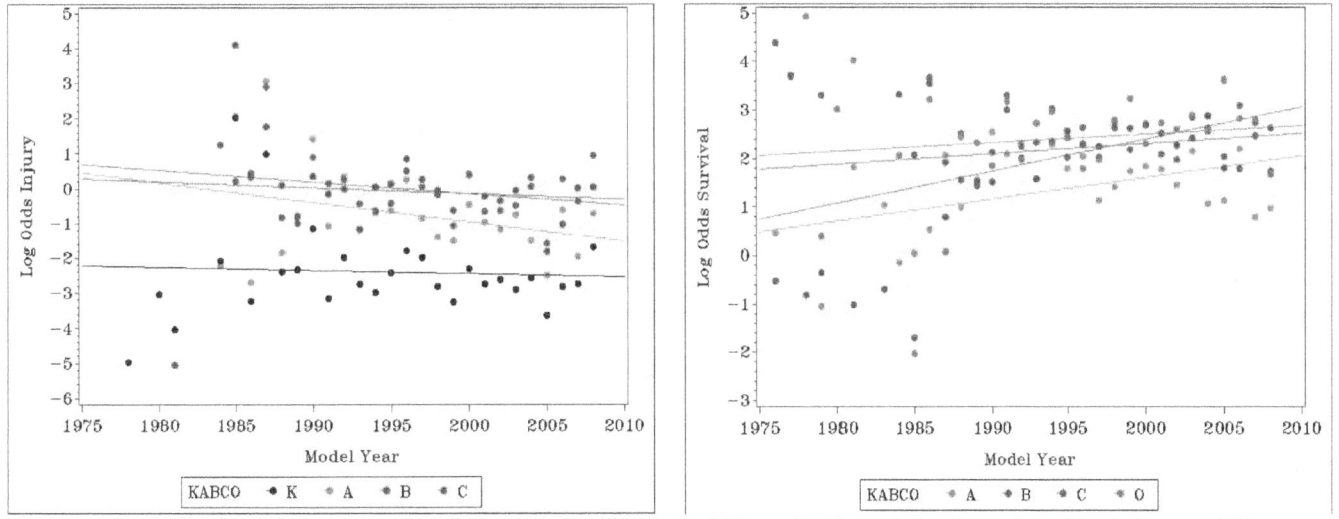

Figure 5-16: Log Odds Injury and Log Odds Survival for Belted 25- to 65-Year-Old Women in LTV Rollovers With Sober Drivers in 2000-2008 (Raw Estimates With Least-Squares Linear Fits)

Figure 5-17 indicates we should also reject the cumulative logistic model, since the slopes clearly appear to depend on KABCO.

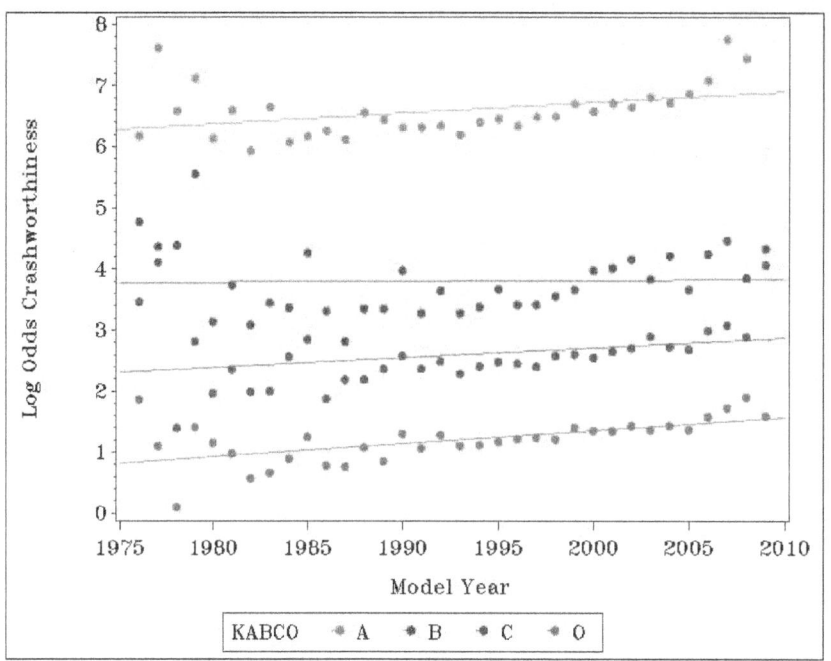

Figure 5-17: Log Odds Crashworthiness for Belted 25-to 65-Year-Old Women in Near Side Car Crashes With Sober Drivers in 2000-2008 (Raw Estimates)

Thus we arrive at the model form:

$$\text{log-odds } P(\text{Injury} \leq k) = g_k + h_k MY$$

where $k \in \{O, C, B, A\}$ and g_k, h_k are linear combinations of effects involving our six categorical variables CT, VT, DA, RU, AC, and G. (Throughout this paper, we use the shorthand MY, DA, RU, AC, and G for the model year, driver alcohol, restraint use, (occupant) age category, and gender factors, respectively, in addition to our shorthand CT and VT for crash type and vehicle type.)

We can fit a model of such a form by fitting four logistic regression models, one for each of KABCO O, C, B, and A, to each of the five imputations of our crash data.[71] In assessing various candidate models in the following, we determine significance by averaging p-values from the imputations. The significant effects for the model for one KABCO level might be different from those for another KABCO level. We use SAS's PROC SURVEYLOGISTIC to fit these models in order to reflect the GES sample design underlying the training data.

Model Training Data
We exclude from the model training data occupants coded in our crash database as having died prior to the crash,[72] as they provide no information on crashworthiness.

We note from Figure 5-18 that the model year 2009 data doesn't fit the pattern, possibly because the estimates have large sampling errors.[73] The pre-1985 model year data sometimes looks less stable than model year 1985 and newer (Figure 5-19), maybe because the sampling errors are large. Including data from any of these model years, whether 2009 or pre-1985, would run the risk of throwing off our model estimates, and so we also exclude them from the training data. No other outliers are apparent.

Thus, our training data for the crashworthiness model consists of occupants in crashes of model year 1985-2008 passenger vehicles in calendar years 2000-2008, excluding those occupants who died prior to the crash.

---

[71] Although there are in a sense five models, we can also form a single crashworthiness model by averaging the predicted values from the five imputation-specific models. Thus we alternatively refer to the crashworthiness *model* or *models*, depending on the context.
[72] Although not traditionally part of the KABCO scale, FARS and GES code as KABCO P (died prior to crash) cases where the police accident report states that a person died prior to the crash or "indicates the person dies as a result of natural causes (e.g., heart attack), disease, drug overdose or alcohol poisoning." (National Highway Traffic Safety Administration, 2008)
[73] There are relatively few crashes of model year 2009 vehicles in our crash data, as the most recent crashes in it occurred in 2008. Consequently, it could be that crashworthiness in model year 2009 vehicles fits the pattern of the earlier model years, and that the error we incurred from sampling a relatively small number of crashes of such vehicles has thrown the estimates off the pattern.

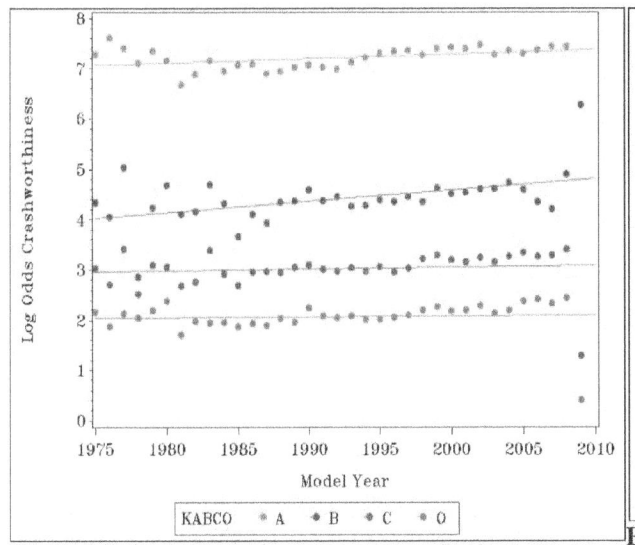

Figure 5-18: Log Odds Crashworthiness for Belted 25- to 65-Year-Old Men in LTV Frontal Crashes With Sober Drivers in 2000-2008 (Raw Estimates With Linear Fits)

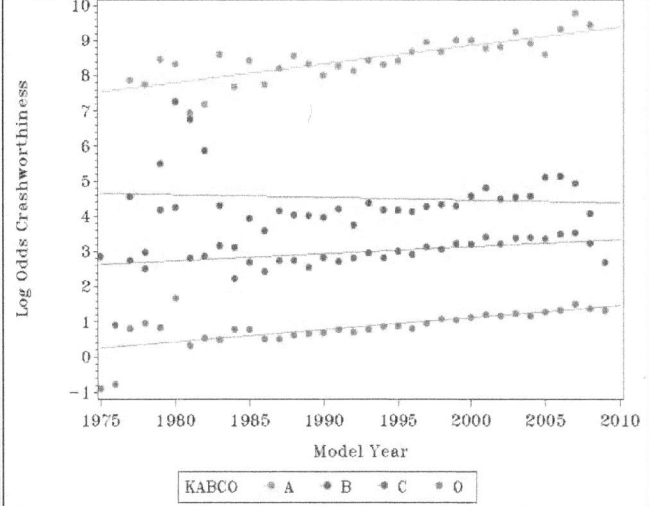

Figure 5-19: Log Odds Crashworthiness for Belted 25- to 65-Year-Old Women in Car Crashes Other Than Frontal, Rollover, and Side, With Sober Drivers in 2000-2008 (Raw Estimates With Linear Fits)

Crashworthiness Model Zero

With six categorical predictors (CT, VT, DA, RU, AC, and G), there are too many higher-order effects to intuit selection from exploratory data analysis. Thus, we start with a basic model (Model Zero) that includes all "main effect" interactions (i.e., CT, VT, DA, RU, AC, G)[74] and slopes (i.e. MY, MY*CT, MY*VT, MY*DA, MY*RU, MY*AC, MY*G)[75]:

log-odds $P(Injury \leq k) \sim$ CT, VT, DA, RU, AC, G, MY, MY*CT, MY*VT, MY*DA, MY*RU, MY*AC, MY*G for k = O, C, B, A.

The Type III tests indicate that most of these effects are significant in all four KABCO models. MY*G and MY*VT are not significant in any KABCO, although MY*VT is marginally significant ($p < 0.1$) for KABCO C. Model Year is significant in the two KABCO levels in which MY*RU isn't.

Table 5-6: Type III Results for Crashworthiness Model Zero[76]

Effect	Whether the Effect Is Significant for the Given KABCO Level (Yes=Significant, No=Not Significant)				# of KABCO Levels in Which the Effect Is Significant
	O	C	B	A	
VT, CT, RU, DA, AC, G, MY*CT, MY*DA, MY*AC	Yes	Yes	Yes	Yes	4
MY	Yes	Yes	Yes	No	3
MY*RU	Yes	No	No	Yes	2
MY*VT, MY*G	No	No	No	No	0

Many of the crashworthiness predictions from Model Zero look quite good (Figure 5-20), although some indicate that interactions could improve them (Figure 5-21). In these and subsequent figures in this (crashworthiness model) section, the lines are the model predictions (not least-squares lines).

---

[74] A priori, one might worry that including both DA and RU might introduce a collinearity problem, since belt use is less common in the presence of driver alcohol. However, our assessment of the final crashworthiness model will allay this concern.

[75] The car-LTV distribution has shifted over the years, with increases in LTV sales in the 1990s (Summers, Hollowell, & Prasad, 2003). The shift in the car-LTV makeup of the vehicle fleet over time (i.e., over model year) should be reflected in the MY*VT term (if crashworthiness differs between cars and LTVs).

[76] Table 5-6 uses a significance level of five percent. We use the same significance threshold (5%) throughout this paper.

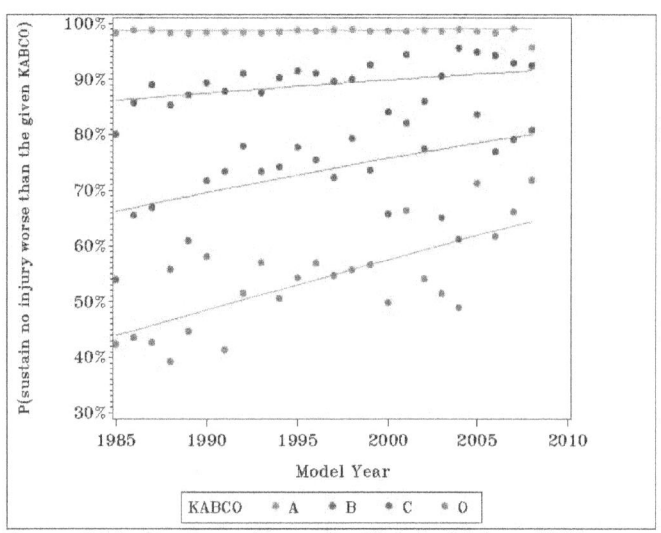

Figure 5-20: Crashworthiness for Unbelted 25- to 65-Year-Old Women in Frontal Car Crashes With Sober Drivers in 2000-2008 (Raw and Model Zero Estimates)

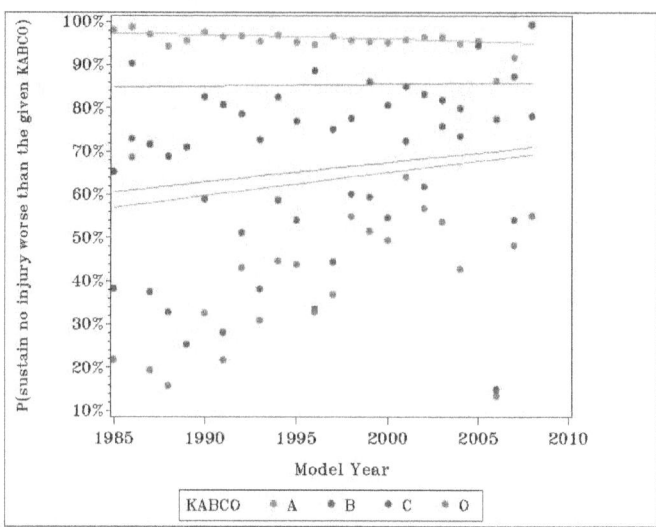

Figure 5-21: Crashworthiness for Unrestrained 14- to 24-Year-Old Males in Frontal LTV Crashes With Non-Sober Drivers in 2000-2008 (Raw and Model Zero Estimates)

Assessing Model Performance

To assess the performance of Model Zero, and to compare its performance with that of other candidate models, we examine how well it predicts our raw crashworthiness estimates, at least those on which we trained the model (i.e., occupants in crashes of model year 1985-2008 passenger vehicles, excluding those who died prior to the crash), with a particular interest in the predictions when the raw crashworthiness estimate is based on enough data to seem pretty reliable (a characterization we will come to quantify).

With two vehicle types, five crash types, three categories of driver alcohol, two categories of restraint use, four occupant age categories, two genders, four non-fatal KABCO levels, and 24 model years (1985-2008) in our model training data, there are potentially 46,080 (= $2 \times 5 \times 3 \times 2 \times 4 \times 2 \times 4 \times 24$) crashworthiness estimates that we could estimate from our 2000-2008 crash data[77] (of crashes in the calendar years 2000-2008). However, our crash data does not permit estimation in all 46,080 cells: Recall that the way we have chosen to incorporate multiple imputations in crashworthiness requires that there be occupants at or below the given injury threshold *and* occupants above the injury threshold.[78] Among the 46,080 cells that would potentially have crashworthiness estimates, 12,556 fail this test.[79] In these cells, our data lacks the information we require to estimate crashworthiness. In the remaining 33,524 cells, we can generate crashworthiness estimates. That is, we have 33,524 raw crashworthiness estimates from the training data on which to assess the performance of any given model. For reference, we shall refer to these 33,524 combinations of crash type, driver alcohol, restraint use, occupant age, gender, model year, and KABCO on which we trained the crashworthiness model as the 33,524 *crashworthiness training cells*. For instance, "belted 25- to 69-year-old women sustaining incapacitating injuries in frontal crashes of model year 1995 cars with sober drivers" is a training cell.

---

[77] Recall that we dropped the year of the crash as a predictor of crashworthiness. Table 4-1 lists the values of our remaining eight predictors.

[78] For instance, if we wish to estimate the ability of model year 2004 cars, when driven by drinking drivers, to protect unrestrained girls under the age of 14 in frontal crashes from non-incapacitating injuries, our definition of crashworthiness from Section 4.2 requires each imputation of our crash database have at least one such girl with at most a non-incapacitating injury and at least one with a more severe injury (i.e., an incapacitating or fatal injury). See Section 4.2 for a complete discussion of our chosen manner of incorporating the multiple imputations in estimating crashworthiness and why we rejected other choices.

[79] Failure can occur in two ways. It could be that there are no occupants of the given vehicle type, crash type, driver alcohol, restraint use, occupant age category, gender, and vehicle model year in some imputation. For instance, none of the five imputations of the 2000-2008 crash data gave rise to any boys under the age of 14 in rollovers of model year 1993 driverless cars. Or, it could be that there are occupants of the desired type in each imputation, but in some imputation, they all lie on one side of the injury threshold. An example of the latter occurs with unrestrained girls under the age of 14 in frontal crashes of model year 2004 cars with drinking drivers: In the first imputation of the crash data, all such girls were either killed or sustained incapacitating injuries (see Table 5-7). Thus, our definition of crashworthiness does not permit estimating the ability of model year 2004 cars, when driven by drinking drivers, to protect such girls against non-incapacitating injury in frontal crashes.

Thinking in 33,524-dimensional space, we are interested in three vectors for a given crashworthiness model (such as Model Zero):

- One "raw crashworthiness" vector: This is the vector of the 33,524 raw crashworthiness estimates from the 33,524 training cells. We shall denote this vector by **RawCW**
- One "model crashworthiness" vector: We have one vector of the 33,524 predictions under the given crashworthiness model. We will denote this vector by **ModelCW**.
- One "residual" vector: This is the vector defined as **Resid** = **RawCW** – **ModelCW**.

We would like to assess the degree to which the residual vector **Resid** is "small", at least relative to the "size" of the raw estimate vector **RawCW**. A natural means to do this would seem to be to generalize the notion of relative error, defining the relative error for the given crashworthiness model as:

$$\text{relative error} = |\textbf{Resid}|/|\textbf{RawCW}|$$

By this measure, Model Zero has a relative error of 19 percent. The 33,524-dimensional error vector **Resid** has length 24, compared to a length of 128 for the raw estimate vector **RawCW**.

However, not all the 33,524 raw crashworthiness estimates in the vector **RawCW** are equally reliable. This is illustrated by the case of unrestrained girls under the age of 14 in frontal crashes of model year 2004 cars with drinking drivers, which appear in our combined FARS-GES database with the following counts.

Table 5-7: Unrestrained Girls Under the Age of 14 Years in Frontal Crashes of Model Year 2004 Cars With Drinking Drivers in 2000-2008 FARS and GES (Sample Sizes)

KABCO Level	Number of Occupants in the 2000-2008 FARS and GES Databases Injured at the Given KABCO Level, by Hotdeck Imputation				
	Imputation 1	*Imputation 2*	*Imputation 3*	*Imputation 4*	*Imputation 5*
O	0	0	0	0	0
C	0	0	0	1	0
B	0	1	0	0	0
A	1	1	2	3	2
K	2	1	1	1	2
Total	3	3	3	5	4

Of the four non-fatal KABCO levels, only one, namely KABCO *A*, defines a crashworthiness training cell.[80] But even for KABCO *A*, estimating crashworthiness would only be based on data from 3-5 girls per imputation, and thus would not seem to produce a reliable figure. Ideally we would like to have some satisfactorily non-small number of occupants contributing to the numerator and the denominator of a raw crashworthiness estimate in each of the five imputations.[81]

Among the 33,524 crashworthiness training cells, 5,314 of them have at least 50 occupants contributing to the numerator and at least 50 to the denominator of the raw crashworthiness estimate in each of the five hotdeck imputations. We shall refer these 5,314 training cells as the *core training cells*. This is illustrated by the case of restrained girls under the age of 14 in frontal crashes of model year 2004 cars with sober drivers.

---

[80] As noted earlier, the first imputation lacks occupants at or below the given injury threshold for the KABCO levels *O*, *C*, and *B*.
[81] The reason we require each imputation to meet a minimum threshold of occupants, rather than applying a threshold to the pooled imputations, stems from the form of our crashworthiness estimator. Recall that we estimate crashworthiness using geometric means that treat each of the five imputations equally. Thus, we would like *each* of the five imputations to contain "sufficient" data before considering a given crashworthiness estimate reliable.

Table 5-8: Restrained Girls Under the Age of 14 in Frontal Crashes of Model Year 2004 Cars With Sober Drivers in 2000-2008 FARS and GES (Sample Sizes)

KABCO Level	Number of Occupants in the 2000-2008 FARS and GES Databases Injured at the Given KABCO Level, by Hotdeck Imputation					Core Training Cell?
	Imputation 1	Imputation 2	Imputation 3	Imputation 4	Imputation 5	
O	285	281	276	279	279	Yes
C	61	61	66	62	60	Yes
B	36	38	36	36	35	No
A	25	23	22	22	24	No
K	13	15	15	13	15	NA[82]
Total	420	418	415	412	413	

KABCO level *O* for such girls (i.e., the training cell comprising restrained girls under the age of 14 who escaped uninjured in a frontal crash of a model year 2004 car with a sober driver) is a core training cell as at least 276 girls contribute to the numerator of the crashworthiness estimate (i.e., escaped injury) in each imputation and at least 133 girls contributed to the denominator (i.e., were injured) in each imputation.[83]

In contrast, KABCO level *B* for such girls is not a core training cell. While the minimum threshold is met for the numerator of the crashworthiness estimate (i.e., the numbers of girls sustaining at most a "possible" injury is at least 50 in each imputation), the denominator threshold is not met (in fact, failing in each imputation). The imputation that comes closest to meeting the denominator threshold is the fifth imputation, in which 29 girls suffered an incapacitating injury or were killed.[84]

If we perform our relative error computation in 5,314-dimensional space, using the 5,314 core training cells, Model Zero has a relative error of 4%. (When restricted to the 5,314 core training cells, the length of the error vector **Resid** decreases to 3, and that for the raw estimator vector **RawCW** decreases to 65.)

As a measure of bias, 35 percent of the 33,524 residuals in **Resid** are positive, compared to 47% of the 5,314 core residuals (residuals in core training cells), indicating the absence of an overall over- or under-estimation of crashworthiness in Model Zero.

Table 5-9: The Performance of Crashworthiness Model Zero

Scope	Relative Error	Percent of Positive Residuals
the 33,524 crashworthiness training cells	19%	35%
the 5,314 core training cells	4%	47%

With nothing else to compare it to, it is difficult to assess the performance of Model Zero. A relative error of 4 percent on the core training cells and 47 percent positive core residuals seem acceptable, but it would seem reasonable to see if we can improve this performance appreciably without fitting to the vagaries of the raw data. We will now add more effects to the crashworthiness model and see how much we can improve this performance.

Crashworthiness Model One
With 15 slopes and 15 intercepts arising from the pairwise interactions of our 6 categorical variables to potentially add as second-order slopes and intercepts, we choose in Model One to add those among these that make intuitive sense to use, namely CT*VT, CT*RU, DA*RU, MY*CT*VT, and MY*CT*RU.

We add CT*VT on the supposition that cars may be more or less crashworthy than LTVs in certain crash modes. For instance, near side impacts of cars would include some collisions of the driver's side of a car with the front of an LTV, which are generally more injurious than a corresponding LTV-LTV impact (Summers, Hollowell, & Prasad, 2003). It is also possible that the increased mass of LTVs protects occupants better than cars do in non-rollover collisions.

---

[82] Our definition of a core training cell is only applicable to non-fatal KABCO levels, since at worst one dies, and so we have put a value of "NA" in this entry of Table 5-8.
[83] This minimum occurs in the fourth imputation (133 = 412 − 279).
[84] We note that the numbers of girls killed varies among the imputations (from 13 to 15). This is not due to unknown injuries being imputed as fatal injuries. We do not allow unknown injuries to be imputed as fatal, on the presumption that fatalities are almost certainly reported as known deaths. Rather, the variation in the numbers of girls killed stems from imputations of FARS cases in which the crash type, vehicle type, driver alcohol, age category, gender, or model year was unknown.

We add CT*RU on the supposition that seat belts and child restraints are more effective in certain crash modes. For instance, the raw data indicate that seat belts are particularly effective in rollovers.

We add DA*RU on the hypothesis that drinking drivers might get into more severe crashes than sober drivers (e.g., drunk drivers speed more often than sober drivers (National Highway Traffic Safety Administration, 2008)) and that in more severe crashes, it is particularly important to be belted.

We add MY*CT*VT, on the supposition that improvement in crashworthiness might differ by VT*CT combination (e.g., maybe improvements for car near side are greater than for LTV nearside). But we won't be shocked if this effect is found not to be significant.

We add MY*CT*RU, in case the improvements to crashworthiness differ by CT*RU combination. This would arise if, e.g., newer vehicles better protect both belted and unbelted occupants in frontal crashes, and better protect belted occupants in rollovers, but little gain has been made in protecting unbelted occupants in rollovers. But as with MY*CT*VT, we won't be shocked if this effect is found non-significant.

Although MY*VT and MY*G were not significant in Model Zero, we retain them in Model One in case they gain significance in the presence of the new terms.[85] Thus, Model One is as follows:

$$\text{log-odds } P(Injury \leq k) \sim \text{CT, VT, DA, RU, AC, G, CT*VT, CT*RU, DA*RU,}$$
$$\text{MY, MY*CT, MY*VT, MY*DA, MY*RU, MY*AC, MY*G, MY*CT*VT, MY*CT*RU for } k = O, C, B, A.$$

Some of the new effects are significant[86] (CT*VT, CT*RU, and DA*RU) in Model One, and some are not (MY*CT*VT and MY*CT*RU). Most non-significant effects are marginally significant ($p < 0.1$) in at least one KABCO level.[87] All Model Zero terms remain significant in Model One, although some are significant in fewer KABCO levels, apparently due to the inclusion of the newly added terms.

Table 5-10: Type III Results for Crashworthiness Model One

Effect	Whether the Effect Is Significant for the Given KABCO Level (Yes=Significant, No=Not Significant)				# of KABCO Levels in Which the Effect Is Significant
	O	C	B	A	
CT, RU, DA, AC, G, CT*RU, MY*DA, MY*AC	Yes	Yes	Yes	Yes	4
VT	Yes	Yes	No	Yes	3
CT*VT, MY*CT	No	Yes	Yes	Yes	3
MY	Yes	Yes	Yes	No	3
DA*RU, MY*RU	Yes	No	No	Yes	2
MY*VT, MY*G, MY*CT*VT, MY*CT*RU	No	No	No	No	0

Model One seems to improve the visual fit of the data, but not tremendously (Figure 5-22).

---

[85] In Model Zero, some effects, such as MY, were significant for some KABCO levels and not others. So one could also consider refining Model Zero by dropping such effects in the KABCO levels in which they were not significant. We do not do this, as retaining such effects in all KABCO levels does no harm and we find it conceptually cleaner.
[86] That is, significant in at least one KABCO level.
[87] In Model One, MY*VT and MY*CT*VT are marginally significant for KABCO A, and MY*CT*RU for KABCO B and A. In contrast, MY*G is not marginally significant in any KABCO level.

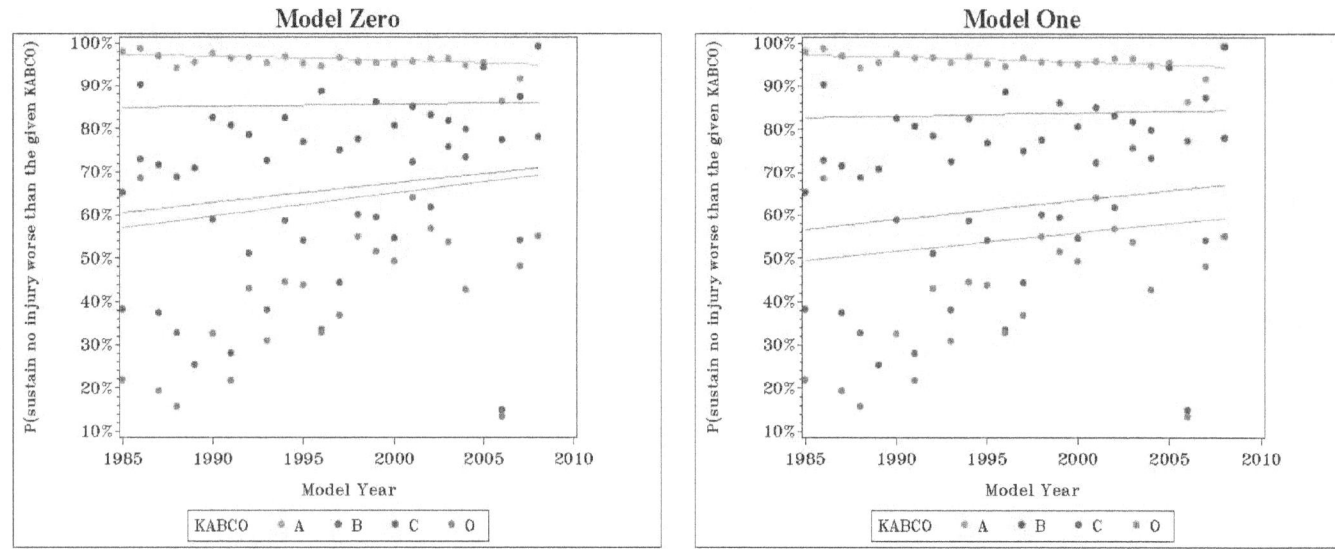

Figure 5-22: Crashworthiness for Unrestrained 14- to 24-Year-Old Males in Frontal LTV Crashes in 2000-2008 With Non-Sober Drivers (Raw, Model Zero, and Model One Estimates)

Model One's performance does not seem markedly better than Model Zero's (Table 5-11).[88]

Table 5-11: The Performance of Crashworthiness Models Zero and One

Scope	Relative Error		Percent of Positive Residuals	
	Model Zero	Model One	Model Zero	Model One
the 33,524 crashworthiness training cells	19%	19%	35%	35%
the 5,314 core training cells	4%	4%	47%	50%

Model Two, the Final Crashworthiness Model
At this point, we could consider adding more terms to the model and/or dropping non-significant ones. No additional terms seem intuitively meaningful to us as impacting crashworthiness or improvements to it, and so we will define our final model by dropping the sole term that was not at least marginally significant in Model One (namely, MY*G).

Thus, our final crashworthiness model (Model Two) is as follows:

The Final Crashworthiness Model

log-odds $P(Injury \leq k) \sim$ CT, VT, DA, RU, AC, G, CT*VT, CT*RU, DA*RU, MY, MY*CT, MY*VT, MY*DA, MY*RU, MY*AC, MY*CT*VT, MY*CT*RU for $k =$ O, C, B, A.

The significance pattern for Model Two is the same as for Model One (Table 5-12) (i.e., no significance status changed with the dropping of MY*G). As in Model One, MY*VT and MY*CT*VT are marginally significant for KABCO A, and MY*CT*RU for KABCO B and A.

---

[88] Although the relative error of Model One does not appear to be any smaller than that for Model Zero in Table 5-11, it is. E.g., on the core training cells, the relative error is 4.4589 percent for Model Zero and 4.1342 percent for Model One, a 7-percent decrease.

Table 5-12: Type III Results for the Final Crashworthiness Model

Effect	Whether the Effect Is Significant for the Given KABCO Level (Yes=Significant, No=Not Significant)				# of KABCO Levels in Which the Effect Is Significant
	O	C	B	A	
CT, RU, DA, AC, G, CT*RU, MY*DA, MY*AC	Yes	Yes	Yes	Yes	4
VT	Yes	Yes	No	Yes	3
CT*VT, MY*CT	No	Yes	Yes	Yes	3
MY	Yes	Yes	Yes	No	3
DA*RU, MY*RU	Yes	No	No	Yes	2
MY*VT, MY*CT*VT, MY*CT*RU	No	No	No	No	0

The visual fits of Model Two are not perceptibly different from Model One (Figure 5-23).

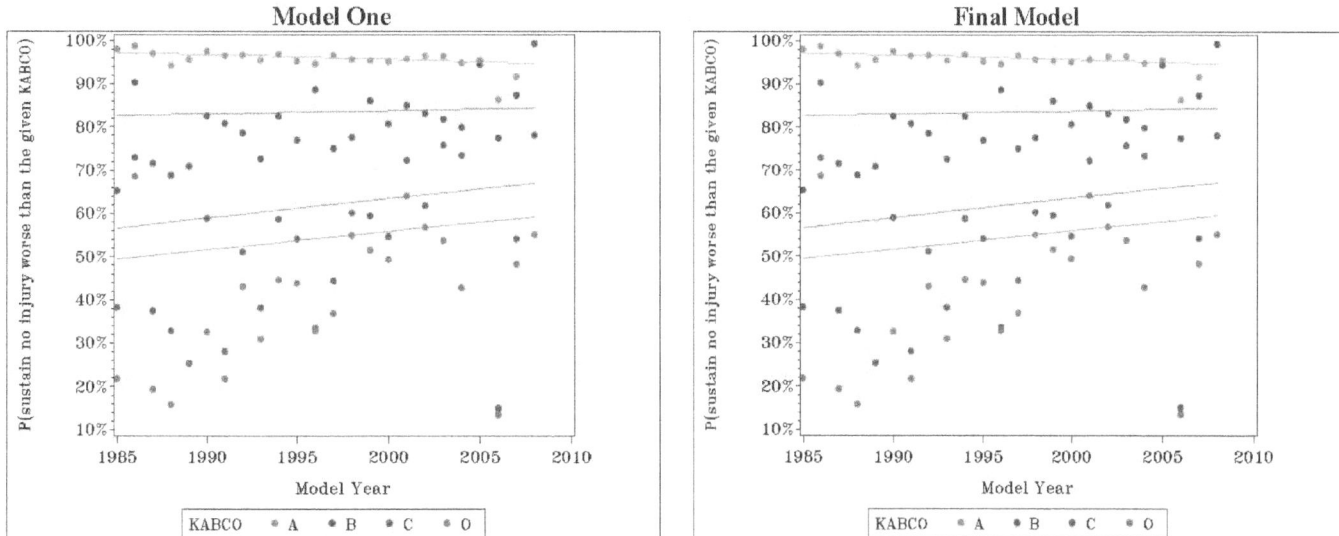

Figure 5-23: Crashworthiness for Unrestrained 14- to 24-Year-Old Males in Frontal LTV Crashes in 2000-2008 With Non-Sober Drivers (Raw, Model One, and Final Model Estimates)

Table 5-13 depicts the relative errors and percent of positive residuals to additional digits and shows the improvement in these quantities in our model development. Overall, the final crashworthiness model decreases the relative error on the core training cells by 7 percent from Model Zero.

Table 5-13: The Performance of Crashworthiness Models Zero, One, and the Final Crashworthiness Model

Model	Performance on the 33,524 Crashworthiness Training Cells				Performance on the 5,314 Core Training Cells			
	Relative Error	% Change from Previous Model	Percent of Positive Residuals	% Change from Previous Model	Relative Error	% Change from Previous Model	Percent of Positive Residuals	% Change from Previous Model
Model Zero	18.9543%	-	35.1122%	-	4.4597%	-	47.2149%	-
Model One	18.7388%	-1%	34.9272%	-0.5%	4.1346%	-7%	49.5295%	5%
Final Model	18.7386%	-0.001%	34.8735%	-0.2%	4.1337%	-0.02%	49.4166%	-0.3%

Figures 5-24 and 5-25 depict the parameter estimates for the crashworthiness model. A complete listing of all 172 parameter estimates for the (final) crashworthiness model appears in the Appendix.

Figure 5-24 depicts the parameter estimates for the effects that do not involve model year. Using unrestrained 25- to 65-year-old women in model year 2000 cars with sober drivers in frontal crashes as the reference group, Figure 5-24 plots the multiplicative effects on the odds of sustaining, at worst, a given level of injury. The reference group's odds of sustaining an injury of at most KABCO k are: 1.2 for $k=$ no injury (O), 3.0 for possible injury (C), 9.3 for non-incapacitating injury (B), and 83.2 for incapacitating injury (A).

For instance, restraint use improves the odds of a 25- to 65-year-old woman surviving a frontal crash in a model year 2000 car more than 11-fold (a multiplicative effect of 11.5), and this is statistically significant. Likewise restraint use improves the odds of such a woman escaping with at most a non-incapacitating injury by more than five-fold and her odds of escaping uninjured by more than four-fold. Restraint use is by far the dominant factor in your injury outcome regardless of your age, gender, type of vehicle, and type of crash.

The model indicates that all else being equal, women fare better than men when it comes to surviving a crash, and men fare better than women when it comes to escaping either uninjured or with a non-incapacitating injury. LTV occupants fare better than car occupants (although the difference between the vehicle types for KABCO "B" is not significant), and rollovers are worse than frontal crashes.

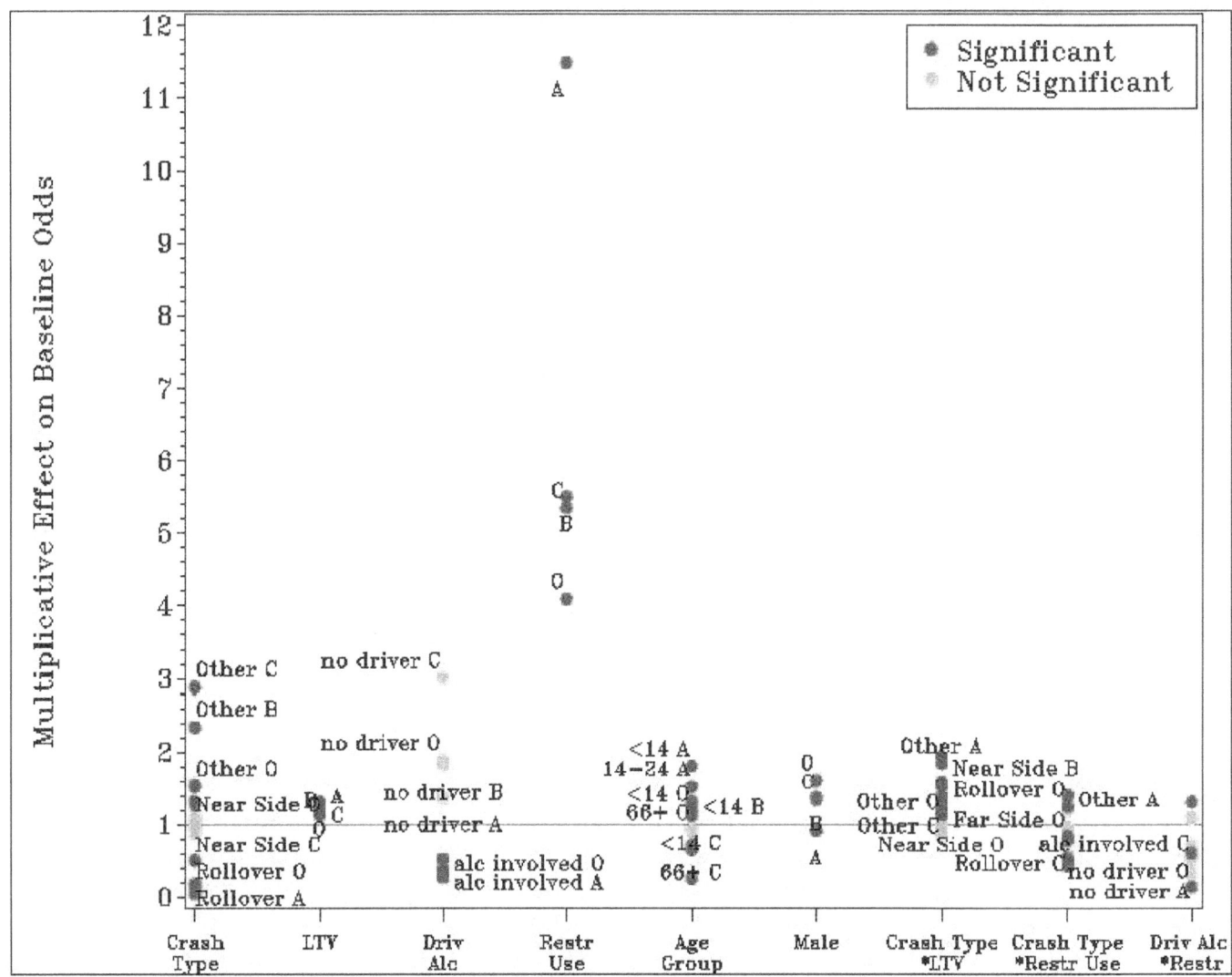

Figure 5-24: The Crashworthiness Model Parameter Estimates That Do Not Involve Model Year

Figure 5-25 depicts the parameter estimates for the effects that involve model year. Namely, it plots the multiplicative effect on the injury odds per unit increase in model year. In our reference group, these multiplicative effects are: 1.039 for KABCO level O, 1.036 for C, 1.028 for B, and 1.006 for A. That is, for unrestrained 25- to 65-year-old women in frontal car crashes with sober drivers, being in a model year 2008 car instead of a model year 2000 car increases the odds of escaping uninjured by a factor of 1.039^8, or about 1.4.

Figure 5-25 indicates that the crashworthiness improvements in LTVs over the modeled period (model years 1985-2008) are not significantly different from those in cars.

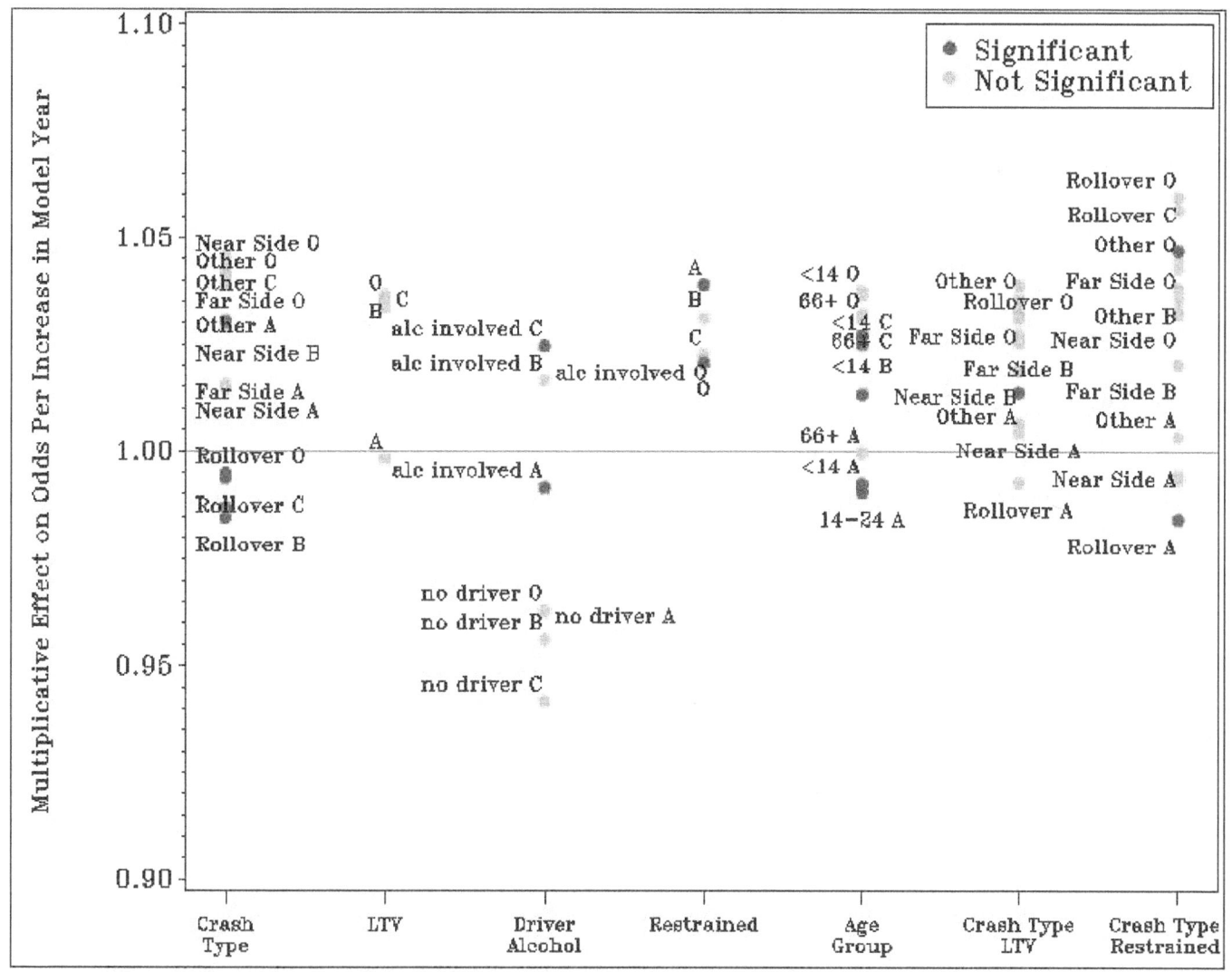

Figure 5-25, The Crashworthiness Model's Effects per Unit Increase in Model Year from Model Year 2000

The blue dots below the horizontal reference line in Figure 5-25 (e.g., for rollovers with KABCO level A) may at first appear to give reason for concern. (Although it is not clear from the point labels in Figure 5-25, the rollover estimates below the reference line are 0.995, 0.994, 0.985, and 0.987 for KABCO O, C, B, and A, respectively, and of these only the KABCO O estimate is not significant.) There are (at least) two reasons why such dots do not necessarily indicate decreased crashworthiness performance.

One possible reason has to do with improvements in crash avoidance. Rollovers might be distinct among crash types in that a rollover that is avoided (whether through Electronic Stability Control or other means) might often result in a crash of a different type (e.g., a frontal crash). In contrast, avoiding a frontal or side impact crash might usually mean avoiding crashing entirely. In improving crash avoidance for rollovers, the remaining rollover crashes may be more severe, leading to an appearance that vehicles may have become less rollover-crashworthy in some scenarios, when they may in fact be protecting us better.

Another reason is that other effects will counteract such an otherwise worrisome blue dot (below the reference line) outside the reference group. For instance, the blue dot with a multiplicative effect of 0.987 for rollovers with KABCO level A applies to the reference group of *unbelted* 25-65 year old women in cars. For belted women of the same age group, cars have *increased* the odds of survival by 1 percent per model year (i.e. the multiplicative effect per model year is (0.987) (1.039) (0.984) = 1.01).

Additionally, the large residuals for rollovers in Figure 5-26 give us reason not to trust the model's predictions for rollovers, and points to potential model refinements. Figure 5-26 plots the difference between the model and raw estimates of crash-worthiness for the various combinations of crash type, vehicle type, driver alcohol, restraint use, occupant age category, gender, and KABCO level, limiting to those combinations in which there are at least 50 sampled occupants contributing to the numerator and denominator of the raw estimate.

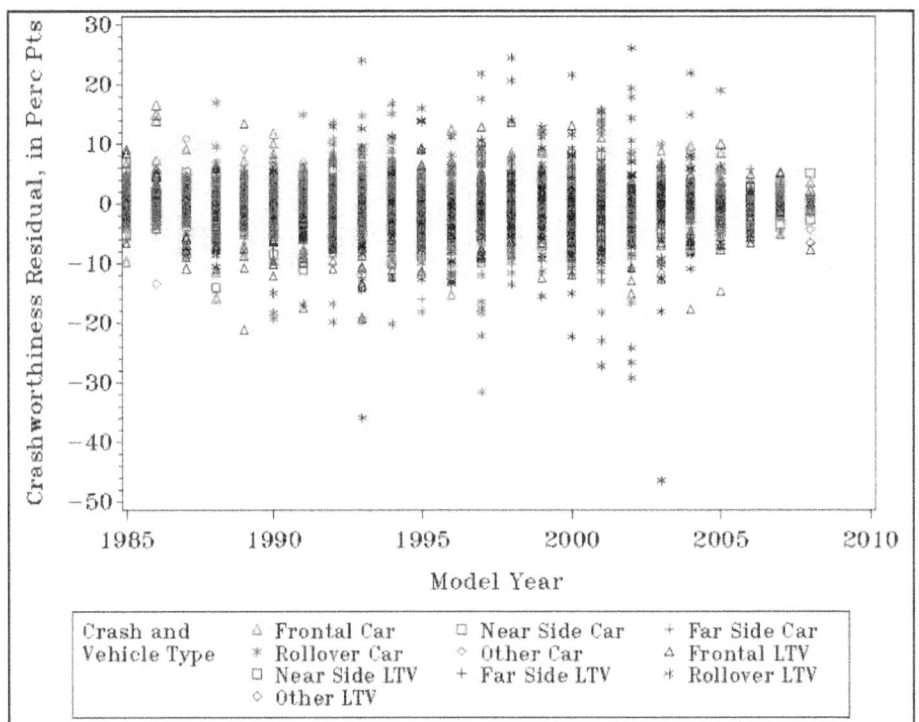

Figure 5-26: Crashworthiness Model Residuals for Cells in Which At Least 50 Sampled Occupants Contribute to the Numerator and Denominator of the Raw Estimate

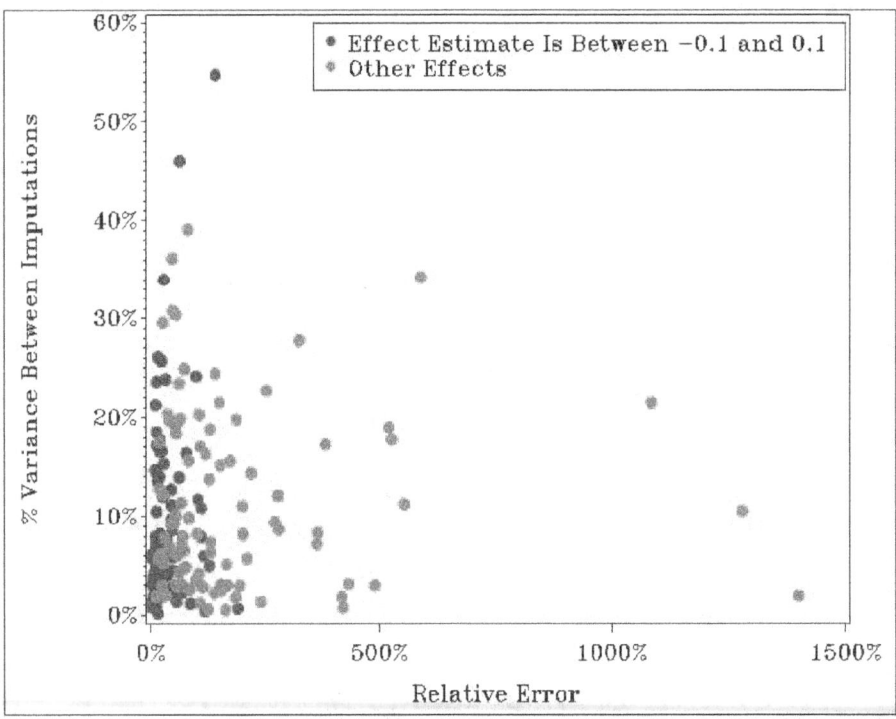

Figure 5-27: Crashworthiness Model Variances

Assessing the sources of variation for our model's parameter estimates, some of the relative errors are quite large (see Figure 5-27). However among parameter estimates that, expressed as linear effects on the injury log-odds, are at least 0.1 in absolute value, the relative error is rarely more than 20%. Imputation accounts for a greater share of the parameter estimates' variances than we saw in the crash avoidance model. Most of the effects with more than 20% of the variance occurring between imputations involve Driver Alcohol, which is difficult to impute accurately. All together though, we do not see evidence of multicollinearity. (Not depicted in

Figure 5-27 are Far Side for KABCO B and MY*(Near Side) for KABCO B, whose relative errors are quite large (both over 1,500%), but these parameter estimates are quite small, with additive effects on the log-odds of injury of –0.00051 and –0.00004, respectively.)

Adequacy of Controls and Potential Future Refinements

How well has our modeling isolated the vehicle component to crashworthiness from the human and environmental components? Our model includes factors for restraint use, age, gender, and driver alcohol, which in combination seek to control for the human element. Whether these factors are sufficient to fully describe all person-related effects on crashworthiness, we will never know. It is conceivable to us that our factors might be adequate, and we cannot think of other factors (on which we have data) that would improve this control.

Our model does not contain any environmental controls. (Recall that calendar year was not significant.) In hindsight, it might well have been beneficial to include factors such as weather and road conditions at the time of crash (which are available in our crash databases), and this could be a possible future improvement to the model.

We also note that our model does not incorporate in any strong way the severity of the crash, such as the change in velocity that the vehicle undergoes during a collision, with the driver alcohol variable only possibly providing a mild surrogate.[89] We are interested in redoing the analysis using the Crashworthiness Data System to incorporate a better measure of crash severity. It would also be desirable to include a better measure of the location of the point of impact relative to the occupant in a non-rollover crash, other than simply the categorization of frontal, near side, far side, and other. It could be particularly helpful to include information on incompatibilities in vehicle-to-vehicle crashes, such as when the bumper of an LTV meets a car driver's window, and incorporate injury type (e.g. leg injuries, whiplash cases, etc).

Other possible future refinements to the model would be to incorporate ages and restraint types (e.g., forward-facing child seat) for children to improve our analysis for this occupant group, and to refine the vehicle classes (e.g., small versus mid-size cars).

5.3 Limitations of Both Models

Our analysis of crashworthiness and crash avoidance is based on police-reported data. Thus, our crash avoidance estimates do not reflect the capacity to avoid the types of minor crashes that tend not to be reported to the police. Likewise the necessary exclusion of non-police-reported crashes means that our crashworthiness estimates underestimate the extent to which vehicles protect their occupants in general crashes.

Having imputed the crash data, our choice of hotdeck imputation cells may well influence which effects show up as significant in the models. For instance, having imputed vehicle type using imputation cells defined by crash year, our data is more likely to indicate a relationship between vehicle type and crash year, diluting the marginal information offered by vehicle type, for a model that involves crash year. Perhaps more dangerously, imputing injury severity from crash type and restraint use means that our crashworthiness model might overstate the impacts of these predictors. However, we feel that the bias reduction likely achieved by including sensible imputation cells outweighs the possible detriment to the statistical models.

To the extent that our models are limited by their assumptions, we list key assumptions underlying the models:

- The effect of calendar year on the log number of crashes per mile driven is linear, and the effect of vehicle age is quadratic.
- Calendar year and vehicle age have no effect on crashworthiness.

These assumptions are supported by Figures 5-2 and 5-14, together with the statistical tests referenced in their discussions. However because they played key roles in developing the mathematical forms taken by our models, they are worth re-iterating. In particular, the second assumption implies that reductions we see in injury rates (such as the improved chance of survival for belted 25- to 65 –year-old women with sober drivers in frontal crashes of model year 2008 cars, compared to model year 2000) must derive from vehicle improvements, and are not due to, e.g., vehicles getting safer or less safe as they age, or particular calendar years being safer than others.

---

[89] presupposing that impaired drivers are in more severe crashes, which might not be true

6. How Much Safer are Newer Vehicles?

As indicated in the introductory sections, the objective of our study is to quantify how much safer newer vehicles are, controlling for human and environmental factors. With our models in hand, we are now in a position to address these questions.

We start by assessing improvements from a personal standpoint – how less likely am I to crash? How much safer am I if I do crash? We then extrapolate to the collective societal benefit – how many fewer crashes occurred as a result of vehicle improvements? How many lives were saved and how many injuries mitigated? How many more could have been prevented/saved/mitigated if older vehicles had been as safe as today's vehicles? This chapter presents the former (improvements from the individual's perspective), while the next chapter presents the latter (the societal perspective).

6.1 The Reduced Likelihood of Crashing

How much less likely are we to crash (if at all) driving a model year 2008 car or light truck than a model year 2000 one?

Recall that our crash avoidance model has factors Vehicle Age and Calendar Year, and the difference of these two (Calendar Year minus Vehicle Age) gives the model year of a vehicle. Thus we could in theory assess the impact of an eight-year (or other) advancement in model year by holding Vehicle Age fixed and advancing Calendar Year by eight years, or by holding Calendar Year fixed and decreasing Vehicle Age by eight years. Which approach to take is a question of which better filters out human and environmental factors. To the extent that drivers of vehicles of a given age drive similarly and under similar environmental conditions,[90] and that environmental changes from year to year are negligible, the first approach better isolates the contribution of vehicles and equipment. To the extent that vehicles driven in the same calendar year are driven under similar conditions and differences in the way vehicles of different ages are driven are negligible, the second approach is better.

The first approach makes more sense to us and is supported by the data. Recall from Section 5.1 that the data indicates that vehicles of different ages are driven differently and/or under different environmental conditions. (See Figure 6-1 below, which we have reproduced from Chapter 5.) Thus, we will conduct our calculations using the first approach, holding Vehicle Age constant and advancing the Calendar Year to assess the impact of newer vehicle technologies, design, and equipment on crash propensity.

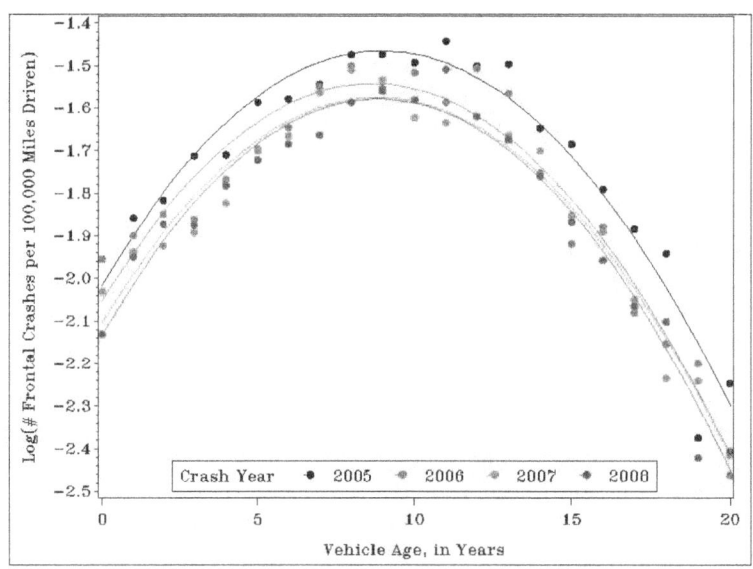

Figure 6-1: Log Crash Rates for Cars in Frontal Crashes, by Vehicle Age and Crash Year (Raw Estimates With Quadratic Fits)

---

[90] If, for instance, vehicles of different ages are driven at night to varying degrees and crashes occur more frequently at night, controlling for miles driven, then some of the differences in crash propensity by vehicle age could be due to nighttime driving, which could be thought of as an environmental condition. (Alternatively, one could view nighttime driving as a human factor, to the extent that one views the propensity to drive at night as a human attribute.)

Having decided how to apply our model, we return to the question of how much less likely we are to crash in a model year 2008 vehicle than a model year 2000 one.

Figure 6-2 depicts the likelihood of crashing in 100,000 miles of driving from our statistical model for cars and LTVs of various model years.[91]

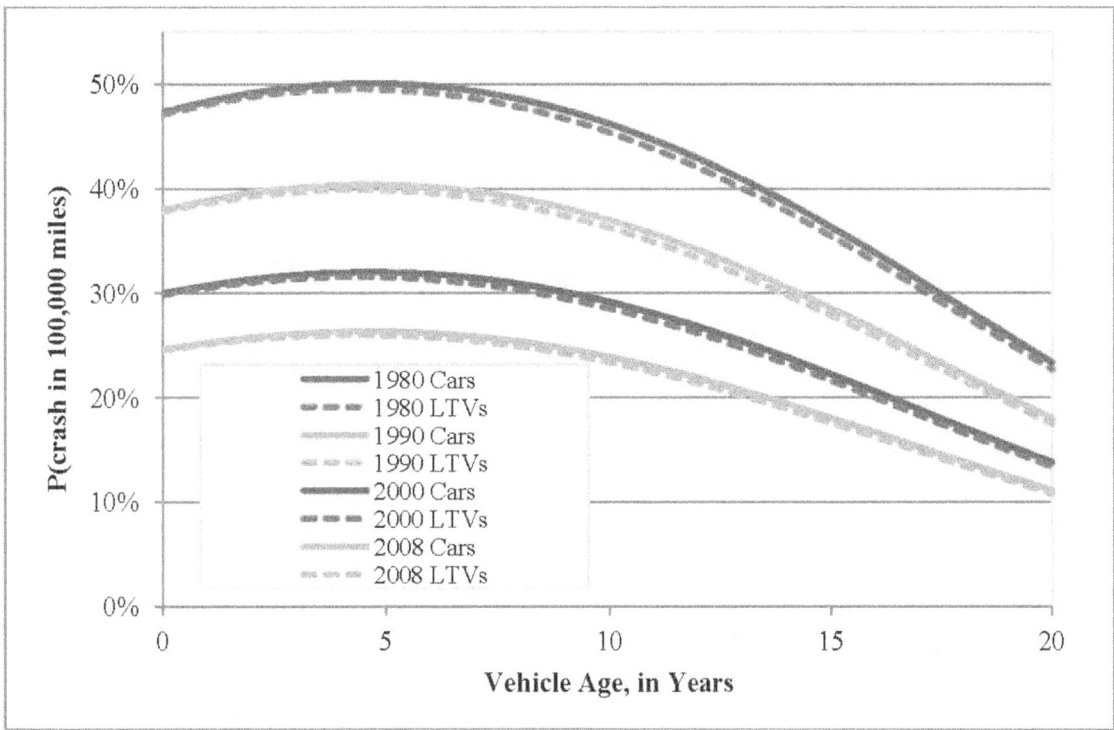

Figure 6-2: Estimated Crash Likelihoods for Cars and LTVs of Various Model Years

According to our model, your chance of crashing in 100,000 miles of driving a model year 2008 car "as new" (i.e., vehicle age 0) is five percentage points lower than it was for a model year 2000 car (30% for model year 2008, 25% for model year 2000).[92] The result is similar for LTVs, and in both cases, the effect of the safety improvements diminishes slightly as the car or LTV gets older and shifts (generally speaking) to different cohorts of drivers or the vehicle is possibly less well maintained. When a car or LTV is 20 years old, the difference in crash likelihoods between a model year 2000 and model year 2008 car is down slightly to 2-3 percentage points.

Our model suggests that crash avoidance has vastly improved between model years 1980 and 2008, with a 22-percentage-point decrease in the crash likelihood for cars driven as new. This may well be stretching our model beyond its predictive range, considering the data on which the model was trained. It would be interesting to compare the model year 1980 estimates in Figure 6-2 to the corresponding raw estimates.[93]

We remind the reader that these results are based on police-reported crashes. The likelihood of getting into a crash of any kind, reported or not, would be higher than the figures presented here. Recall also the limited ability to control for human factors in our

---

[91] Figure 6-2 depicts the estimates predicted by our crash avoidance model. It would arguably be better to compute baseline estimates from the raw data (e.g., the raw crash avoidance estimates for model year 2000 cars and LTVs at age 0) and apply the change in this estimate predicted by the model, in order to account for any systemic overprediction or underprediction by the model. However, our model's residuals (as depicted in Figure 5-12) indicate no systemic over- or under-prediction and so we do not apply such baselining.
[92] Preliminary results from this paper were presented at the 2011 Conference on the Enhanced Safety of Vehicles (Glassbrenner, 2011). At that time, we estimated the likelihood of crashing in a model year 2000 car driven for 100,000 miles as new to be 33 percent, and that for model year 2008 to be 27 percent, a 7-percentage-point reduction (difference due to rounding). We subsequently determined that these figures were based on a model trained on data that included vehicles that were over 20 years old (which, as indicated in Section 5.1, was contrary to our intention).
[93] In this report, we limited to crash data from the 2000-2008 calendar years in order to access the GES General Area of Damage variable, which first appeared in the 2000 data file. This variable is not needed to compute the likelihood of getting into a crash of *any* type, and thus one could compute raw estimates of this likelihood for ,e.g., model year 1980 cars at ages 0-19 years (the age 20 estimate having already been computed by us using the calendar year 2000 crash files). We have not done so.

crash avoidance model, and the reader should ultimately decide the extent to which s/he believes estimates, such as those in Figure 6-2, based on such a model. In our judgment, our estimates are based on about the best available techniques applied to about the best available data, but we will never know if they are correct or the extent of their error.

The Age at Which a Vehicle Is Most Crash-Prone
Recalling from Figure 6-1 that crash rates peak when a vehicle reaches nine years, we may wonder why in Figure 6-2, they peak at 4-5 years of age. The answer is that the estimates in these two figures measure different things. Figure 6-1 demonstrates that in each *calendar year*, the highest crash rates are from 9-year-old vehicles. Figure 6-2 depicts that in each *model year*, crash rates peak at 4-5 years of age. The two can be perfectly consistent, since the model year distribution of vehicles on the road fluctuates with the calendar year. In a sense, the fact that the crash rates in Figure 6-1 peak later than those in Figure 6-2 is a further illustration that crash avoidance has improved in newer vehicle fleets: If all fleets were equally crash-prone as they came off the lot, the highest crash rates in a given calendar year would come from 4- to 5–year-old vehicles. The fact that they peak with 9–year-old vehicles says that the reduction in crash propensity for the 4- to 8-year-old vehicles, compared to the 9-year-old vehicles, is large enough to outweigh the driver effect and shift the peak crash rate to the 9-year old vehicles. [94][95]

This phenomenon is also exhibited in our crash avoidance model. Our model form takes the form:

$$\log(\text{crash rate}) = a(x-b)^2 + c(y-2000) + d$$

where x and y denote vehicle age and calendar year, respectively, and the constants a, b, c, and d depend on the type of crash and vehicle under consideration (e.g. frontal crashes of cars). Substituting $y = x + z$ gives the following re-parameterization of our surface in terms of model year (z) and vehicle age (x):

$$\log(\text{crash rate}) = a(x-B)^2 + c(z-2000) + D$$

where $B = 2b - c/a$ and $D = d - 3ab^2 + 4bc - c^2/a$. For a given model year, the peak crash rate occurs at age B, while crash rates in a given calendar year peak at b. With our parameter estimates (which are listed in the Appendix), we have $B < b$, i.e. $b < c/a$.

The Crash Likelihood Surfaces
Figures 6-3 and 6-4 depict the likelihood of crashing at least once in 100,000 miles of driving for cars and LTVs. We present them simply to show the surfaces, although it is difficult to distinguish the car and LTV surfaces.

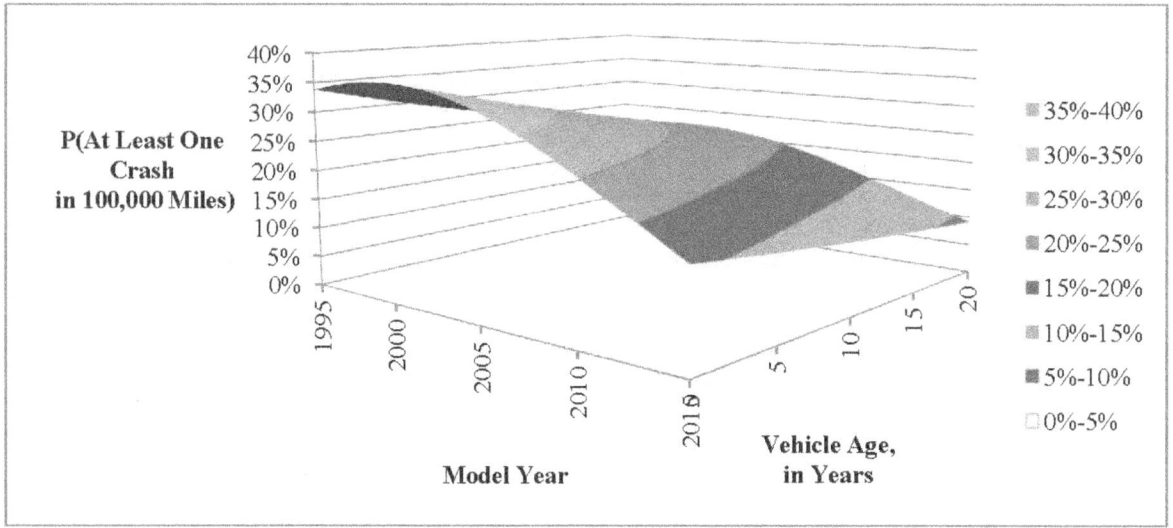

Figure 6-3: The Likelihood of Crashing in 100,000 Miles for Cars

---

[94] One might also note that Figure 6-1 depicts raw estimates and Figure 6-2 depicts model-based estimates, or that Figure 6-1 is for frontal crashes while Figure 6-2 reflects all crash modes. However neither of these is the reason for the nine-year versus 4-5 year discrepancy. The model year-specific raw estimates of the likelihood of a frontal crash also suggest the nine-year peak.
[95] That crash rates peak at vehicle ages 4-5 indicates that the worst (most crash-prone) drivers are people who drive 4-5 year old vehicles.

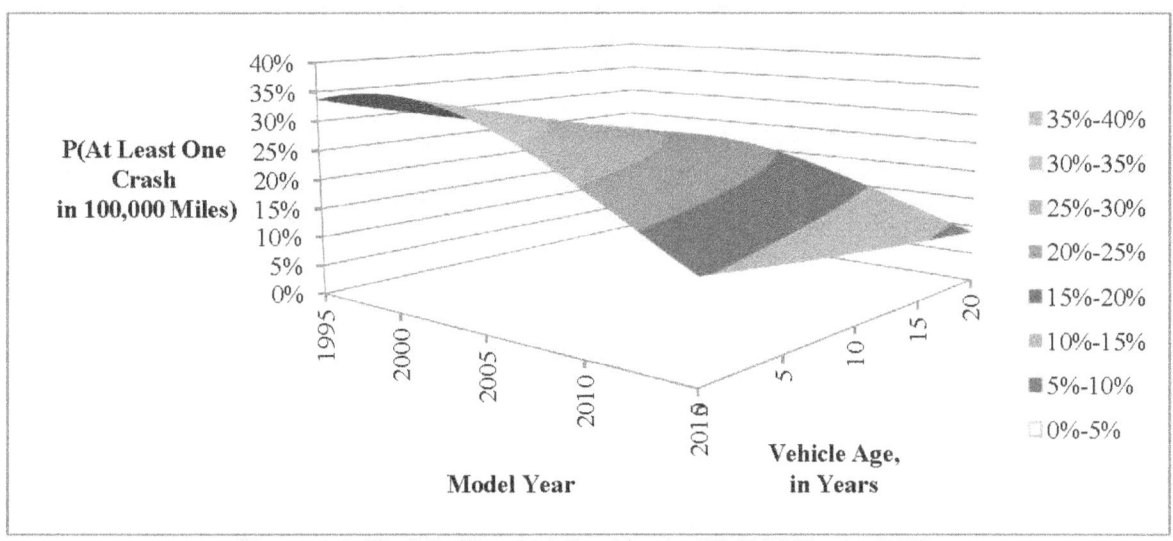

Figure 6-4: The Likelihood of Crashing in 100,000 Miles for LTVs

Isolating Crash Avoidance Improvements
So far we have identified improvements in the capacities of cars and LTVs to avoid crashes of any type. But has the degree of improvements been fairly uniform, or are their particular types of vehicles and crashes that have improved more than others?[96]

For vehicle types, recall that the interaction between Calendar Year and Vehicle Type was found insignificant in our model, indicating the degree to which crash avoidance has improved in cars is similar to that for LTVs.

In which crash modes have the greatest improvements come? For the manner in which we are applying the model, the model terms that reflect improvement are the parameter estimates that involve calendar year (i.e. CY and CY*CT). These give the following percentage reductions in the number of crashes in 100,000 miles of driving resulting from a one-year increase in model year.

Table 6-1: Crash Avoidance Improvements, by Crash Type

Crash Type	The Percentage Reduction in the Number of Crashes of the Given Type in 100,000 Miles Resulting From a One-Year Increase in Model Year
Frontal	3%
Rollover	6%
Side	3%
Other	3%

The reductions in this table apply to both cars and LTVs, as our model does not include an interaction between Calendar Year, Vehicle Type, and Crash Type (which was found non-significant). The reductions also apply to all vehicle ages, recalling that our exploratory data analysis found no interaction between Calendar Year and Vehicle Age.

On a percentage reduction basis, the greatest improvements to crash avoidance have come in rollover prevention. Recalling that the likelihood of crashing is approximately the crash rate,[97] Table 6-1 says that each one-year increase in model year reduces your likelihood of experiencing a rollover, by about six percent. Although rollovers are far less frequent than other crash types, they are the most injurious, and so this six-percent reduction in the occurrence of a rare event has substantial implications for saving lives and mitigating injuries, as we will see in Chapter 7.

The three-percent reductions in the other three crash types (frontal, side, and "other") suggest that vehicles have undergone improvements that reduce crashes of any type by three percent in each new fleet, with additional improvements (perhaps ESC) that reduce rollovers by an additional three percentage points. Note that these are per-fleet reductions, so that a ten-year increase in model

---

[96] While it would be natural to ask the same question for other subpopulations (e.g., whether the improved crash avoidance has benefitted vehicles driven by drunk drivers to a greater or lesser degree than those driven by sober drivers), we unfortunately cannot answer such questions, lacking e.g. data on miles driven drunk. The only subpopulations that the data allow us to assess for crash avoidance improvements are those posed in the question (i.e., those defined by vehicle type and crash type).
[97] Recall from Chapter 3 that $CA \approx exp(CR) \approx 1-CR$, and so the crash likelihood $1-CA$ is approximately equal to CR.

year would reduce non-rollover crashes by nearly a third and rollovers by 60 percent. Figure 6-5 depicts the improvements over an eight-year span, comparing the 2000 and 2008 fleets.

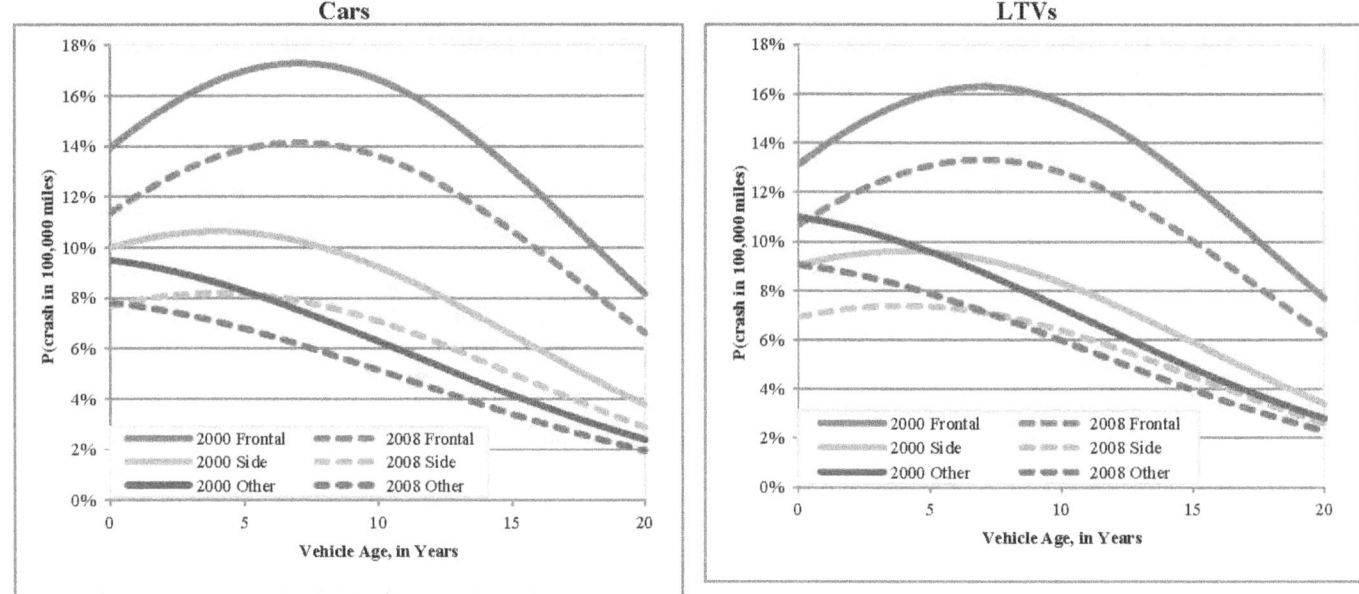

Figure 6-5: Improvements in Avoiding Non-Rollover Crashes, by Crash and Vehicle Type

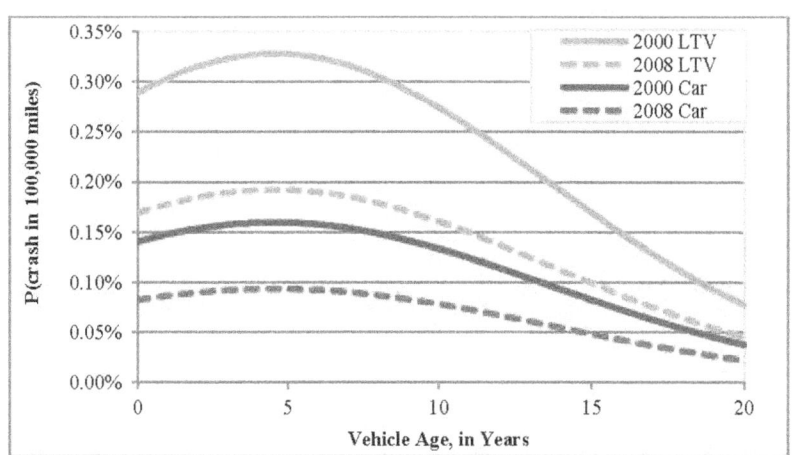

Figure 6-6: Improvements in Avoiding Rollover Crashes, by Vehicle Type

Again, we remind the reader that the estimates in these figures are model-based estimates from police-reported crashes.

6.2 The Reduced Likelihood of Injury in a Crash

If we are unfortunate enough to be in a crash, how much better off are we to crash in a model year 2008 vehicle than a model year 2000 vehicle?

When answering the analogous question for crash avoidance (how much less likely are we to crash), the form of our statistical model dictated that we make a decision regarding how to apply it.[98] Our crashworthiness model, in contrast, poses no such difficulty. Recall that this model has the form:

$$\text{log-odds } P(Injury \leq k) \sim \text{CT, VT, DA, RU, AC, G, CT*VT, CT*RU, DA*RU,} \\ \text{MY, MY*CT, MY*VT, MY*DA, MY*RU, MY*AC, MY*CT*VT, MY*CT*RU} \quad \text{for } k = \text{O, C, B, A.}$$

---

[98] Our crash avoidance model incorporated model year via terms for the calendar year and vehicle age, and so we had to decide whether to assess fleet improvements by advancing the calendar year or reducing the vehicle age. We chose the former.

With model year as the model's sole numeric variable, it is clear that we simply advance the model year to assess improvements in crashworthiness.

We are presented with other challenges to overcome, though, in assessing improvements to crashworthiness. For instance, it would be disingenuous to assess, e.g., improvements to crash survival in cars by plotting estimates of the survival likelihoods that are developed in some way to make them independent of occupant characteristics (gender, age, restraint use, willingness to drive drunk or travel with a drunk driver).[99] This is because, as we discussed in the context of crash avoidance, people who drive (or ride in) 10-year-old vehicles are likely different than those in new vehicles, perhaps generally being from different age groups, genders, or having different propensities to drink and drive.[100] Thus, we find it safest to present crashworthiness only for specific occupant types (as well as vehicle types and model years), and not attempt to generalize them.[101] As the number of occupant types is prohibitively large (with four age groups, two genders, two restraint statuses, and three driver alcohol statuses) to depict the improvements in all, we will depict the improvements in a "baseline group" and then discuss the extent to which the improvements in other groups differ. Our baseline group will consist of belted 25- to 65-year-old women traveling with sober drivers.[102] [103]

Crashworthiness Improvements for Belted 25- to 65-Year-Old Women With Sober Drivers
Figure 6-7 presents the fleets' improved ability to protect the baseline group. If you are a belted 25- to 65-year-old woman with a sober driver and you are in a crash (a police-reported crash of any type), your chance of surviving was already quite high (over 99.5%), even in a model year 1985 vehicle. Thus the fleet improvements for survival are necessarily slight. Your chance of escaping incapacitation in the crash has improved from 97-98 percent for the model year 1985 fleet to 98.5-99 percent for the model year 2008 fleet. Perhaps most notably, your chance of walking away uninjured has improved from 73-78 percent in a model year 1985 vehicle to 82-87 percent in a model year 2008 vehicle, a 9-percentage-point improvement in each vehicle class (car, LTV).[104]

---

[99] One could for instance estimate such a survival likelihood by weighting the survival likelihoods (whether model-predicted or raw estimates) for the various occupant, driver alcohol, and crash types by their relative incidence in crashes.
[100] We have not analyzed the possible existence or strength of these relationships.
[101] Recall that all factors were found significant in the crashworthiness model, and (on a related note) as seen in the Appendix, all factors have a parameter that is significantly different from zero, so there is no clear category over which it would be "safe" to collapse in assessing crashworthiness improvements without running the risk of confounding relationships with vehicle age in a way that ultimately misconstrues a change in crashworthiness figures as a fleet improvement.
[102] A woman in the baseline group may be driving (and sober) or may be a (sober or not) passenger with a sober driver.
[103] In an attempt to avoid confusion, we are intentionally using different terminology here ("baseline group") than we used when alluding to the reference levels of the categorical variables in the crashworthiness model (where we used the term "reference group"). The reference group for the crashworthiness model comprises unbelted 25- to 65-year-old women in frontal crashes of model year 1985 cars with sober drivers. The baseline group we chose to use for assessing crashworthiness improvements consists of belted 25- to 65-year-old women traveling with sober drivers. There is no canonical choice for either, and no reason for them to coincide or share common features.
[104] Our crash data contains 2,000 – 70,000 women in the baseline group (i.e., belted, 25-65 years old, with sober drivers) for each vehicle type (car, LTV) and model year (1985 – 2008). Thus the crashworthiness estimates in Figure 6-7 would not seem to suffer from a data reliability problem.

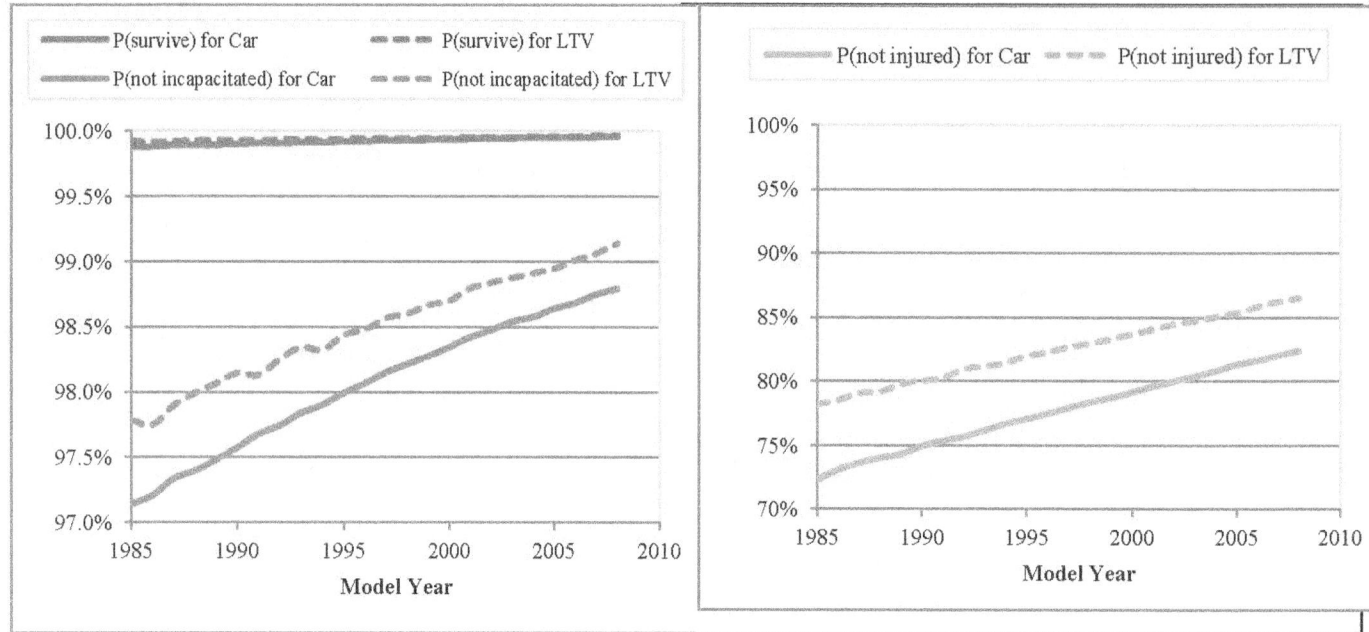

Figure 6-7: Improved Crashworthiness in the Baseline Group

Figures 6-8 through 6-10 detail the results by the type of crash, still focusing on our baseline class of persons to protect. Figure 6-8 presents the improved ability to survive a crash, Figure 6-9 presents the increased protection against incapacitating (or fatal) injuries, and Figure 6-10 presents the improved chance of walking away uninjured.[105] Each of these figures presents the results for rollover crashes on a different y-axis, as rollovers are so much more severe than other crash types.

All figures indicate improvements in crashworthiness, or at least, the lack of decline. The sole indication of a potential decline in crashworthiness (survival in LTV rollovers) is not statistically significant, as we will see in Table 6-2. The most notable improvements in protection against death and incapacitation seem to be in cars for near side and "other" crashes (non-rollover crashes other than frontal and side), in that the curves in these categories climb more steeply than others. With regards to escaping uninjured, the greatest improvements seem to occur in non-rollover crashes other than frontal and side, where both vehicle categories have shown remarkable improvements.

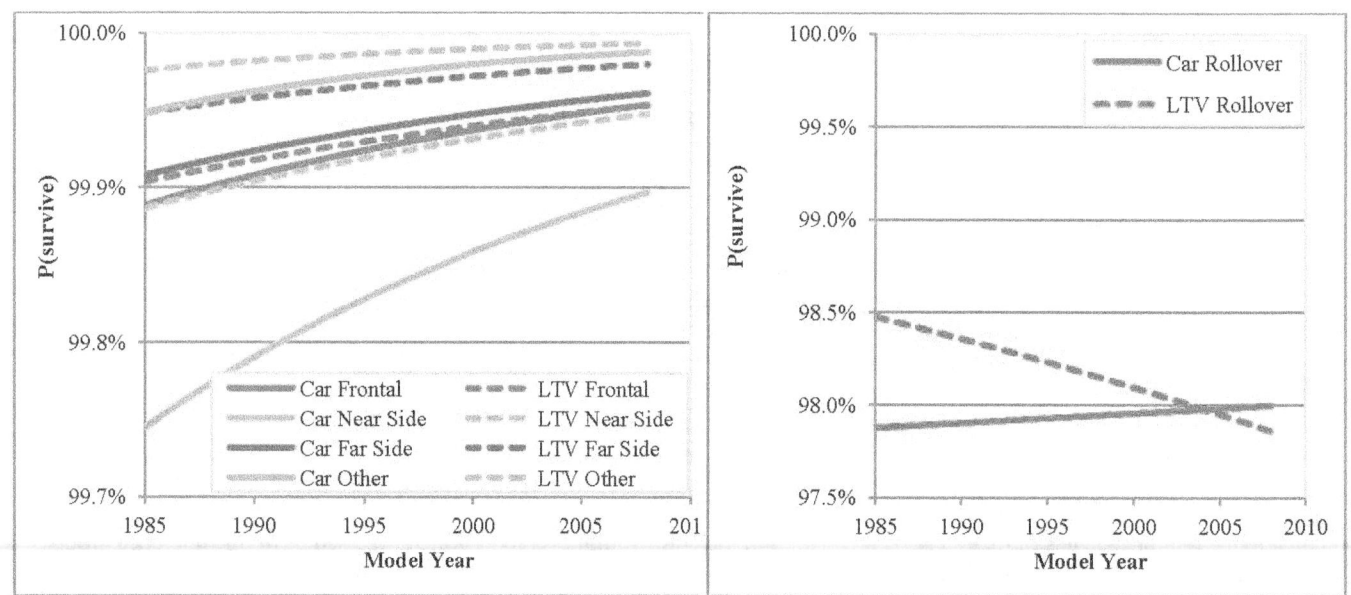

Figure 6-8: Improved Crash Survival in the Baseline Group

---

[105] We do not assess improvements for KABCO level C (possible injury), as we find this less meaningful.

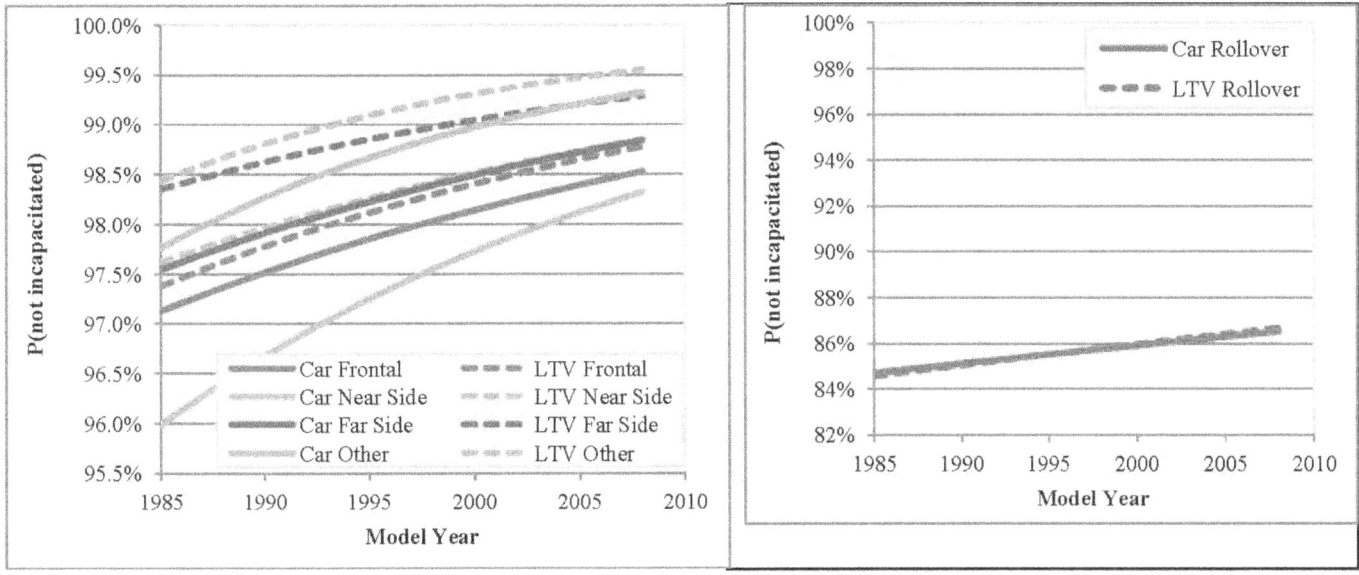

Figure 6-9: Improved Protection Against Incapacitating (or Fatal) Injury in the Baseline Group

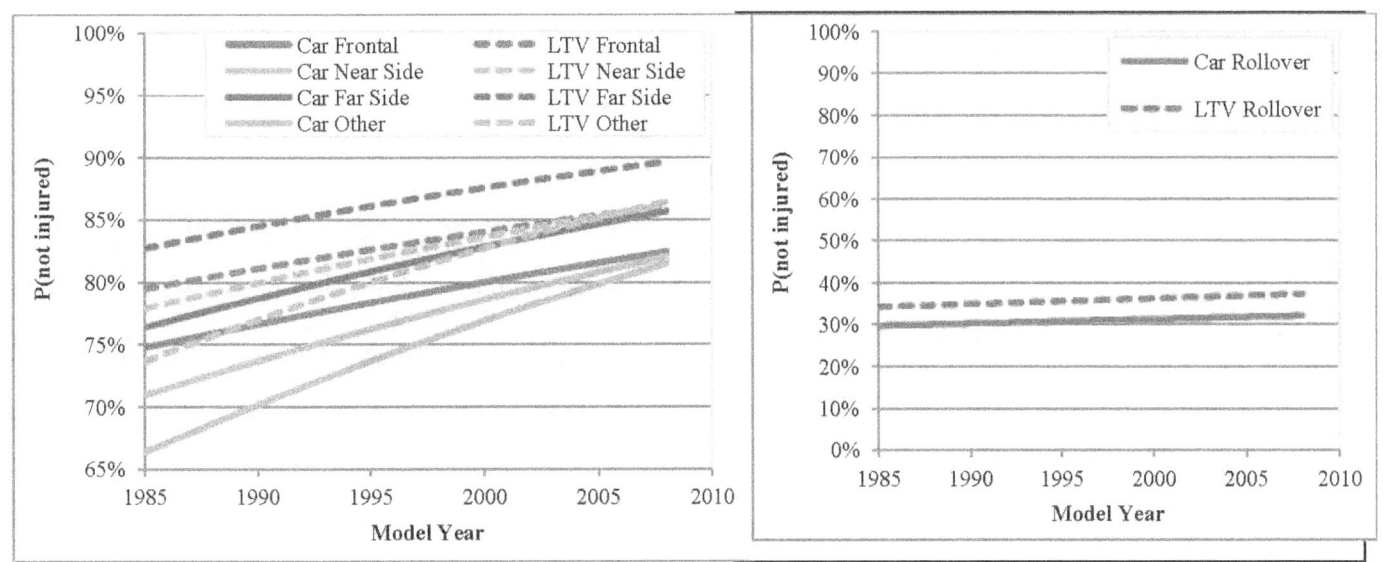

Figure 6-10: Improved Chance of Walking Away Uninjured in the Baseline Group

Which of the improvements indicated in Figures 6-8 through 6-10 are statistically significant? This is answered by examining the statistical significance of the model's parameter estimates. For instance in the crashworthiness model of survival, the parameter estimate for MY*CT(Other) is statistically significant, while that for MY*CT(Far Side) isn't, indicating that the apparent improvements for car "other" crashes in Figure 6-8 is real, while that for car far side crashes might not be real and just reflect sampling and or imputation error.[106,107] Table 6-2 presents the same information for other model terms whose impact is depicted in Figures 6-8 through 6-10.[108] For more easily interpreted figures, we express the parameter estimates in Table 6-2 as the percentage increase to crashworthiness log-odds that would result from a 10-year increment in model year.

---

[106] Our standard errors are computed on our multiply imputed datasets and reflect both sampling and imputation variance.
[107] Again, we caution the reader that our parametrization of the categorical variables in the SAS modeling used unbelted 25- to 65-year-old women in frontal crashes of model year 1985 cars with sober drivers as the reference category. This is not to be confused with the baseline group of belted 25- to 65-year-old women with sober drivers we use to assess improved crashworthiness.
[108] Recall that there are separate parameter estimates for the different KABCO levels (e.g., the parameter MY will have one estimate for each of KABCO's O, C, B, and A).

Table 6-2: Parameter Estimates for MY*RU, MY*CT*RU, MY*CT, MY*VT, and MY*CT*VT, Expressed as a Percentage Increase in Crashworthiness Odds Resulting From a 10-Year Increase in Model Year
[Estimates in highlighted cells are statistically significant with 95% confidence]

Effect		*Percentage Increase in Odds From a 10-Year Increase in Model Year*		
		Survival Odds	Non-Incapacitation Odds	Non-Injury Odds
MY*RU	Restrained	40%	2%	-13%
MY*CT*RU	Far Side	-11%	-8%	1%
	Near Side	-9%	9%	-2%
	Other	-1%	9%	9%
	Rollover	-18%	19%	26%
MY*CT	Far Side	11%	11%	6%
	Near Side	12%	0%	10%
	Other	29%	14%	5%
	Rollover	-16%	-36%	-33%
MY*VT	LTV	-6%	5%	0.2%
MY*CT*VT	Far Side	10%	-4%	-2%
	Near Side	2%	-10%	-2%
	Other	0%	-2%	1%
	Rollover	-11%	-3%	-4%

Note that the some of the terms in Table 6-2 counteract one another. For instance, the improved survival indicated by the positive MY*RU estimate (40% over 10 model years) is enough to counteract the negative MY*CT(Rollover) and MY*CT(Rollover)*RU(Restrained) estimates (-16% and -18% over 10 model years), resulting in improved survival for the baseline group in car rollovers. However, the same positive MY*RU term is not adequate to counteract if we also incorporate the negative terms for MY*VT and MY*CT*VT (-6% and -11%), resulting in the decreasing function in Figure 6-8.[109]

Crashworthiness Improvements for Other Occupant Classes

Improvements to crashworthiness are found by examining the remaining terms that interact with model year, namely: MY, MY*AC, MY*RU, and MY*DA. Table 6-3 presents these estimates.

Table 6-3: Parameter Estimates for MY, MY*AC, and MY*DA, Expressed as a Percentage Increase in Crashworthiness Odds Resulting From a 10-Year Increase in Model Year [Estimates in highlighted cells are statistically significant]

Effect		*Multiplicative Effect on ...*		
		Survival Odds	Non-Incapacitation Odds	Non-Injury Odds
MY		5%	33%	41%
MY*AC	< 14 Years	-12%	-3%	2%
	14-24 Years	-13%	-14%	-7%
	> 65 Years	-5%	-3%	-4%
MY*DA	Driver not sober	-12%	-12%	-10%
	No driver	-35%	-52%	-51%

The significant decreases in the survival odds for children and youths under 25 years of age indicate that the improved protection against death has not extended to these more vulnerable occupants to the degree that is has for 25- to 65-year-old occupants. Treating all occupant classes equally,[110] Table 6-4 presents the percentage changes in three crashworthiness indicators – the likelihood of death,

---

[109] We are not asserting statistical significance here.
[110] That is, the averages in Table 6-4 are straight averages over the occupant categories, treating, e.g., belted and unbelted occupants equally, regardless of their incidence in the population. The intention here is to indicate the areas in which the greatest gains in crashworthiness have occurred, regardless of how commonly such crashes occur. Chapter 7 will address the same question incorporating the relative incidence, by presenting the numbers of occupants whose injuries were prevented or mitigated as a result of vehicle improvements.

incapacitation, or injury – between the model year 2000 and model year 2008 fleets.[111] They indicate that the greatest gains in crashworthiness have come in crashes other than frontal, side, and rollover. Large gains were also made crashes involving sober drivers, for restrained occupants, and for adult occupants (ages 25+).

Table 6-4: Average Reductions in the Likelihood of Death, Incapacitation, and Injury Between the Model Year 2000 and 2008 Fleets, Treating All Occupant Classes Equally

Feature		% Change in P(death)		% Change in P(incapacitated)		% Change in P(injury)	
		Car	LTV	Car	LTV	Car	LTV
Crash Type	Frontal	-4%	-2%	-9%	-16%	-8%	-11%
	Near Side	-11%	-8%	-16%	-13%	-15%	-15%
	Far Side	-9%	-13%	-18%	-18%	-15%	-13%
	Rollover	10%	22%	4%	4%	0%	0%
	Other	-23%	-18%	-23%	-22%	-16%	-17%
Driver Alcohol	Sober driver	-11%	-6%	-17%	-17%	-14%	-14%
	Non-sober driver	-3%	-1%	-10%	-11%	-8%	-8%
	No driver	4%	0%	53%	47%	23%	26%
Restraint Use	Restrained	-15%	-10%	-15%	-15%	-9%	-9%
	Unrestrained	2%	5%	-9%	-11%	-11%	-14%
Occupant Age	< 14 Years	-5%	-3%	-17%	-20%	-14%	-16%
	14-24 Years	-2%	2%	-8%	-5%	-9%	-8%
	25-65 Years	-11%	-8%	-11%	-17%	-9%	-13%
	> 65 Years	-10%	-6%	-15%	-13%	-11%	-9%
Occupant Gender	Female	-7%	-4%	-13%	-13%	-10%	-10%
	Male	-7%	-3%	-12%	-13%	-11%	-12%

We note that the absence of interaction terms involving model year and gender indicates that crashworthiness improvements have been similar for men and women.[112]

As with the crash avoidance results, the crashworthiness improvements presented in this chapter are model-based results based on police-reported crashes. One would expect unreported crashes to generally be less severe and vehicles to keep their occupants more safe in unreported crashes. However because of available data, we have limited our assessment of crashworthiness to police-reported crashes. The extent to which protection in unreported crashes has improved remains an open question.

---

[111] We have not assessed the statistical significance of the results in Table 6-4. Some may indicate actual declines in crashworthiness and others may be a result of sampling and imputation errors that don't indicate actual declines.
[112] The significant gender term in the crashworthiness model indicates that it is better to be male than female in a crash. However the *improvements* to crashworthiness have been similar in each gender.

7. Reductions in Crashes, Injuries, and Fatalities

Having quantified improvements from a personal standpoint (how less likely am I to crash? how much safer am I if I do crash?), we next examine the collective societal benefit – how many fewer crashes occurred as a result of vehicle improvements? How many lives were saved and how many injuries mitigated? How many more could have been prevented/saved/mitigated if older vehicles had been as safe as today's vehicles?

Consider, for example, the collective travel of model year 2001-2009 cars during the calendar year 2008. Suppose we hypothetically transplant their occupants to model year 2000 cars, and, rewinding the clock back to January 1, 2008, submit them to travel the same routes they traveled in 2008. Our crash avoidance model predicts that more crashes would result (than actually occurred), and our crashworthiness model predicts that the injuries resulting from any given crash would be more severe. Applying the crash avoidance model, we can estimate the reduction in the number of vehicle-crashes for each crash type (rollover, frontal, side, and other crashes) and vehicle type (car, LTV). Applying the crashworthiness model, we can estimate the reduction in the number of occupants injured at each (KABCO) level, at least if we assume that the crashes that did not occur have the same numbers of occupants per vehicle as those that did. Through this hypothetical exercise and model application, we thus estimate the reductions in vehicle involvements,[113] injuries,[114] and fatalities resulting from the extent to which model year post-2000 cars are safer than model year 2000 cars.[115]

Through a similar thought experiment, this time placing occupants of older vehicles (say pre-model year 2000 cars) in newer ones (say, model year 2000 cars), we can estimate the vehicle involvements that could have been prevented, lives that could have been saved, and injuries that could have been mitigated had these occupants been in the newer vehicles. (Here, though, we will want to be careful not to apply the models to very old vehicles to which we wouldn't expect the models to apply.)

Notation

We shall use the following shorthand notation throughout this chapter. For $2000 \leq y \leq 2008$ and $2000 \leq m \leq 2008$, we define *Scenario m+* to denote the probabilistic experiment in which we replace each post model year m car that was on the road in calendar year y with a randomly chosen model year m car, do likewise with LTVs, rewind the clock to the beginning of year y, and submit these vehicles to travel the same roads at the same times under the same driving conditions with the same occupants as they did/had in year y.[116][117] Likewise we define *Scenario m–* to denote the experiment in which we replace cars (respectively, LTVs) from the model year 1974, 1975, …, and m-1 fleets[118] with randomly chosen cars (respectively, LTVs) from the model year m fleet, and submit the replacement vehicles to the same travel with the same occupants as the original vehicles underwent in year y. For instance, the examples described in the preceding paragraphs concern Scenario 2000+ and Scenario 2000-.

---

[113] Our crash avoidance model will estimate the reduction in *vehicle involvements* (numbers of vehicles in crashes), not the reduction in the number of *crashes*. Since in some sense, the point is that *crashes* are prevented, we sometimes refer to such (as in the title of this chapter), but our quantifications are denominated in vehicle involvements. More sophisticated modeling would be required to denominate the results in crashes prevented.

[114] As fatal injuries are themselves injuries, we will sometimes use the word "injuries" to include fatal injuries. It should be clear from context whether the inclusion of fatal injuries is intended.

[115] Obviously, this exercise, which we have conducted here using model year 2000 cars, can be applied with any choice of model year and either vehicle type (car or LTV) to assess the benefits of post model year x vehicles being safer than model year x vehicles. In this report, we present results for $2000 \leq x \leq 2008$, assessing the benefits of improvements made to these fleet years.

[116] Our notation suppresses the dependence of the scenarios on the crash year.

[117] We could equally consider scenarios in which $m < 2000$, but presume this to be of less interest.

[118] The choice of 1974 as a cut-off model year is both a practical and intentional limitation. Our registration data from Polk and thus our crash avoidance estimates aggregate figures from pre-1974 model year vehicles. Thus the oldest fleet year to which we could estimate the reduced crash involvement under our thought experiment is the model year 1974 fleet. Although we could generate estimates of the reduced injury outcomes for *any* fleet year, we also do not want to apply our models to fleets far outside the range to which the models were fit.

7.1 Crashes Avoided and Avoidable

Fix a calendar year y between 2000 and 2008, inclusive. We estimate the reduction $VCAvoidScen(m, v, c, y)$ in the number of vehicle-crashes of each vehicle type v (car, LTV) and crash type c (rollover, frontal, side, and other) that would result in year y under Scenario $m+$ as follows:[119]

$$VCAvoidScen(m, v, c, y) = \sum_{k=m+1}^{y+1}(CR(y + m - k, c, v, y - k) - CR(y, c, v, y - k)) VehMiles_{yvk}$$

where $CR(y, c, v, a)$ denotes our model for the crash rate as a function of calendar year, crash type, vehicle type, and vehicle age from Chapter 5 and $VehMiles_{yvk}$ denotes the collective miles[120] driven by model year k vehicles of type v in calendar year y from Chapter 4.[121] That is, we replace the model year k vehicles of the given vehicle type, for $1974 \leq k \leq m$, with model year m vehicles of the same vehicle type and apply the crash rate that the model year m vehicles are predicted to have when they reach *the same vehicle age*.[122] We do this because we saw in Chapter 5 that vehicle age has an appreciable impact on the crash rate. The term $VCAvoidScen(m,v,c,y)$ estimates the number of vehicle crashes that didn't happen in calendar year y because of collective improvements in the model year $m+1$ through 2009 fleets. To isolate the portion attributable to the individual fleets, we simply take successive differences:

$$VCAvoid(m, v, c, y) = VCAvoidScen(m, v, c, y) - VCAvoidScen(m-1, v, c, y)$$

$$= \sum_{k=m+1}^{y+1}(CR(y - k + m, c, v, y - k) - CR(y - k + m - 1, c, v, y - k)) VehMiles_{yvk}$$

This formula makes intuitive sense, as we are comparing the crash rates between the model year m and $m-1$ fleets. The quantity $VCAvoid(m,v,c,y)$ estimates the number of vehicle crashes that didn't happen in year y because of improvements in the model year m fleet, i.e. because of the extent to which model year m vehicles are safer than model year $m-1$.

We similarly estimate the expected increase $VCAvoidableScen(m,v,c,y)$ in the number of vehicle crashes of vehicle type v and crash type c that would result under Scenario $m-$ in year y as[123]:

$$VCAvoidableScen(m, v, c, y) = \sum_{k=1974}^{m-1}(CR(y, c, v, y - k) - CR(y + m - k, c, v, y - k)) VehMiles_{yvk}$$

In Scenario $m-$, the replacement vehicles (which have model year m) are newer than the vehicles they are replacing (i.e., the index k in the summation is less than m). In calculating the avoidable crashes, we apply the crash rate that the model year m vehicles had when they were as new as the vehicles they are replacing (an event that occurred in the calendar year $y+m-k$). Thus $VCAvoidableScen(m,v,c,y)$ estimates the number of vehicle crashes that didn't have to happen in year y because of the collective improvements in the model year m fleet, compared to the fleets from model years $1974,\ldots, m-1$. We likewise isolate the model year m contribution as:

$$VCAvoidable(m, v, c, y) = VCAvoidableScen(m, v, c, y) - VCAvoidableScen(m-1, v, c, y)$$

$$= \sum_{k=1974}^{m-1}(CR(y - k + m - 1, c, v, y - k) - CR(y - k + m, c, v, y - k)) VehMiles_{yvk}$$

---

[119] We do not have information on the distribution of the reduction in crashes (of the various types) under the various realizations of Scenario $m-$ and thus estimate the expected reduction in crashes according to what would occur if one were to replace vehicles (of type v) with model years in the range 1976 to m with a "typical" model year m vehicle.
[120] Of course $VehMiles_{yvk}$ and the (vehicle-)crash rates have to be in comparable units of measure. In our calculations we have denominated each in hundreds of thousands of vehicle miles.
[121] Again, so as not to apply our models much beyond the range of model years to which they were fitted, we make the conservative assumption that no crashes would be avoided by replacing a model year pre-1976 vehicle with a model year m vehicle.
[122] A model year m vehicle will attain age $y-k$ in calendar year $y+m-k$.
[123] We assume that there are few if any vehicles of model year $y+2$ or later on the road in calendar year y.

The term *VCAvoidable(m,v,c,y)* estimates the number of vehicle crashes that didn't need to happen in year *y* because of improvements in the model year *m* fleet over the previous year's fleet.[124]

Calculating these items from our datasets results in the following estimates for the 2008 crash year.[125]

Impact of Vehicle Improvements on the Number of Crashes – the Big Picture
Figure 7-1 presents the reduced numbers of vehicles in crashes in 2008 resulting from improvements to individual model year fleets. That is, this chart shows the estimates $\sum_v \sum_c VCAvoid(m, v, c, 2008)$ in blue and $\sum_v \sum_c VCAvoidable(m, v, c, 2008)$ in orange. (In both cases, *v* runs through the two vehicle types (car, LTV) and *c* through the four crash types.)

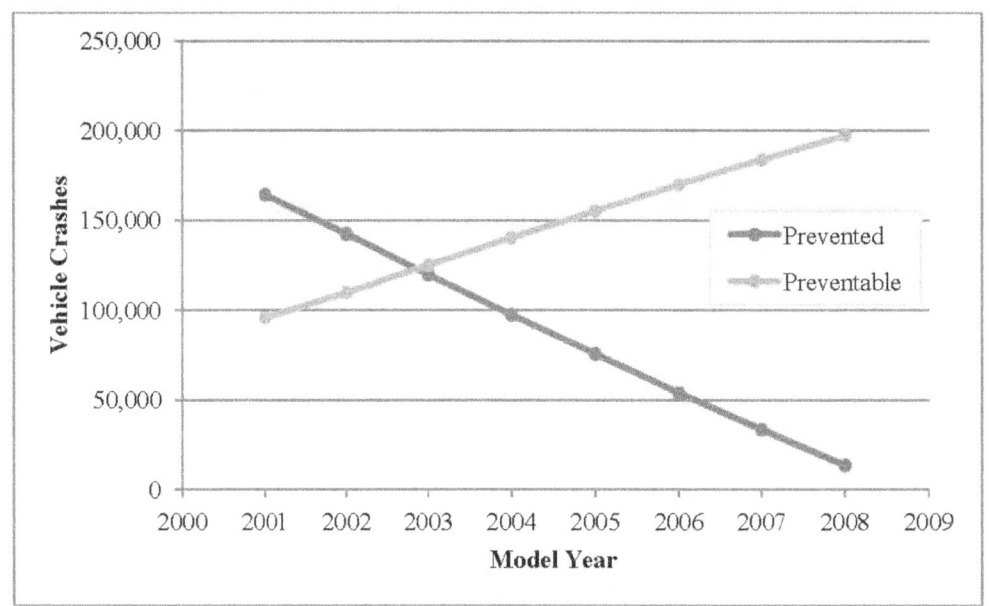

Figure 7-1: Prevented and Preventable Vehicle Crashes in 2008 from Improvements to the Passenger Vehicle Fleet

It is striking that although the collective miles driven increases with model year (and sharply so, recalling Figure 4-2 in Chapter 4), the number of vehicle crashes prevented *decreases* with model year. For instance, model year 2001 cars and LTVs were driven a collective 165 billion miles in 2008 and reduced vehicle involvements by 160,000. In contrast model year 2007 cars and LTVs were driven many more miles (221 billion miles), but prevented far fewer crashes, reducing vehicle involvements by 30,000. This is due to a combination of two phenomena, one of which makes perfect sense and the other being more subtle. The one making perfect sense is that model year 2007 vehicles likely have the safety features of their model year 2001 counterparts, and our measure of improvement attributes the benefit of such features to the model year 2001 fleet. In other words, our measure assesses the marginal benefit and does not attribute to model year 2007 what it can attribute to model year 2001.

The other, more subtle phenomenon goes back to our relationship between crash rate and vehicle age. Recall from Figure 6-1 that the crash propensity increases from purchase to about nine years old and then drops again. Thus, the seven-year-old vehicles in 2008 (model year 2001) crashed more frequently (per mile driven) than the one-year-old vehicles (model year 2007), presenting more opportunity for crash reduction in the older vehicles.

---

[124] Talking about avoidable crashes presents a challenge of language that we don't have for avoided ones. While *VCAvoidScen(m, v, c, y)* and *VCAvoid(m, v, c, y)* are easily distinguished in words (as the numbers of vehicle involvements prevented by improvements made after, or in, a given model year), this is not as easily done for *VCAvoidableScen(m, v, c, y)* and *VCAvoidable(m, v, c, y)*. Both could represent what one means by "vehicle involvements preventable by improvements made in a given model year". (The difference lies in what one compares the performance given fleet to, in one case, comparing to the actual vehicles on the road and in the other to the previous model year's fleet.) In our judgment, the more likely interpretation of "vehicle involvements preventable by improvements made in a given model year" lies with *VCAvoidable(m, v, c, y)*, and so we shall be careful to add a reference to the comparison group (e.g., "vehicle involvements preventable by replacing older vehicles with model year *m* vehicles") when talking about *VCAvoidableScen(m, v, c, y)*.
[125] Recall that we limit our analysis to passenger vehicles in police-reported crashes. Thus our estimated reductions in crash involvements do not include the prevention of a presumably substantial number of crashes that, had they occurred, would not have been reported to the police.

As the prevented crashes decrease with fleet year, the preventable crashes (depicted in orange in Figure 7-1) increase. The collective improvements made to the model year 2001 fleet (improvements, that is, compared to the 2000 fleet) would have prevented an estimated 96,000 vehicle involvements in the calendar year 2008, had every car and LTV on the road possessed them (the fleet improvements). Note here that we once again attribute to the model year 2001 fleet only what we cannot attribute to the 2000 fleet, and so the 96,000-vehicle reduction represents a *marginal* improvement. In comparison, the model year 2007 fleet would have prevented many more crashes, reducing the numbers of cars and LTVs in crashes in 2008 by an estimated 180,000 vehicles. That is, the marginal improvement to the 2007 fleet is nearly twice as large as that for the 2001 fleet. To some extent this would seem only natural (that the later model year's improvements would prevent more crashes), since a larger number of vehicles on the road in 2008 lack the fleet year 2007 improvements than lack the improvements made to the 2001 fleet. However here we have the countervailing force of vehicle age: The vehicles that make up the difference (those having the 2001 fleet improvements but not the 2007 improvements) are about 2-7 years old in 2008 (e.g., many model year 2001-2006 cars[126]). As such, they are driven by drivers who are less crash-prone than the vehicles lacking the 2001 fleet improvements (which are largely over 7 years old), offering less opportunity for crash reduction. But even though they contribute a smaller potential benefit on a per-vehicle basis, they *do* contribute a benefit, making the marginal benefit (in reduced vehicle involvements) from the 2007 fleet improvements larger than that for the 2001 fleet.

Tabulating the savings from Figure 7-1, improvements to the model year 2001-2009 fleets prevented the crashes of 700,000 vehicles in 2008, a remarkable benefit. Had no safety improvements occurred after model year 2000, there would have been roughly 10 million passenger vehicles in crashes in 2008, instead of 9 million. We remind the reader that these figures, and all figures in this chapter, are model-based results based on police-reported crashes.

Crashes – Areas of Greater and Lesser Impact
Based on the individual perspective presented in Chapter 6, we would expect the crash reduction distributed proportionally among the two vehicle types and would expect a disproportionate share of the benefit to occur in rollovers. While Chi-Square tests do detect a difference between the distributions of actual and prevented vehicle crashes, the differences are not where we expect them to be.

Figure 7-2 presents the distribution of the 9 million vehicle crash involvements from 2008 among the eight crash and vehicle type combinations. It also presents the same eight-category distributions for three assessments of crash reduction (the vehicle crashes avoided by improvements to the collective model year 2001-2009 fleets, and those from the individual model year 2001 and 2008 fleets) and two assessments of crash reduction potential (vehicle crashes that could have been avoided by improvements to the model year 2001 and 2008 fleets).[127] [128] As expected the Chi-Square Test finds that the reduction in vehicle crashes from model year 2001-2009 improvements is not proportionate among the eight categories. However the differences are not where we expect: The benefit occurs disproportionately among side impacts and among LTVs. Specifically, 23 percent of the 9 million vehicle involvements in 2008 were in side crashes, but 33 percent of the 700,000 fewer vehicle involvements from model year 2001-2009 improvements would have been in side impacts. Similarly, LTVs accounted for 41 percent of the actual vehicle involvements, but they account for 44 percent of the vehicle crash reduction.

The societal picture doesn't match what we would have expected based on analyzing the individual's perspective because rollovers are so rare. Although the greatest improvements have been in protection against rollover, we don't detect a disproportionate reduction in rollovers because they are so rare.[129]

---

[126] To us, any way in which the model year 2007 fleet is safer than the model year 2006 fleet constitutes a fleet improvement (to the 2007 fleet). Such improvements could stem from new technologies introduced in the 2007 fleet, or the incorporation of a technology that existed in 2006 into a wider share of the 2007 fleet. Our reference to many model year 2001-2006 cars having the fleet year 2001 improvements but not the fleet year 2007 improvements is a simplification made to illustrate the impact of the vehicle age effect.
[127] We use 2008 as the most recent model year fleet in Figure 7-2 because some of the cell sizes for the model year 2009 fleet are too small for the Chi-Square Test.
[128] What appear to be small red and pink lines in the figure indicate rollovers, which are extremely rare among crash types.
[129] It is also possible that vehicle age again confounds the picture, causing a mismatch between the personal and societal pictures.

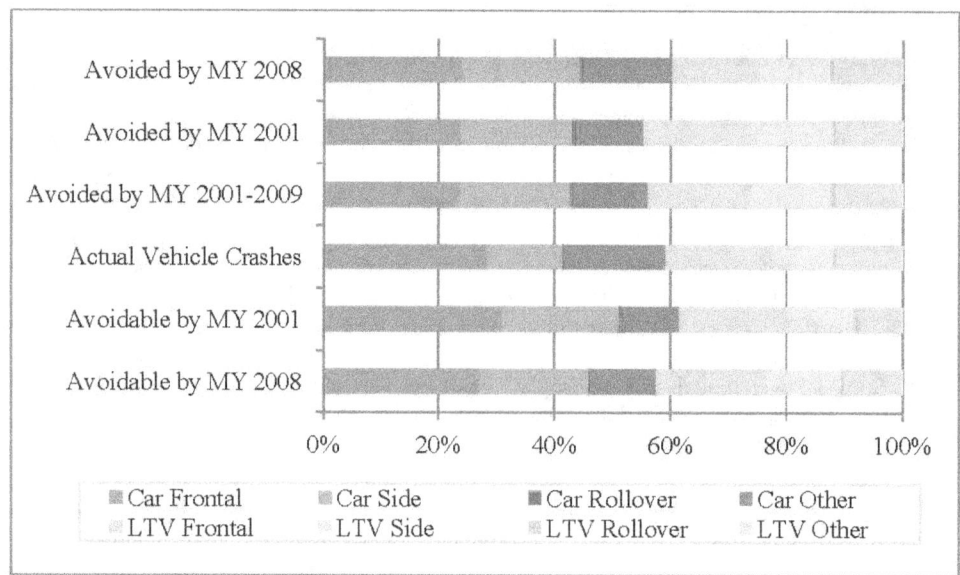

Figure 7-2: The Distribution of Vehicle Crashes in 2008, and of Those Avoided and Avoidable From Fleet Improvements

The distribution of reduced vehicle involvement also differs for the model year 2001 and model year 2008 fleets. A disproportionate share of the crash reduction from model year 2008 improvements occurs for cars and in "other" crashes. For instance, cars account for 60 percent of the vehicle crashes that didn't happen as a result of model year 2008 improvements, compared to 56 percent of those from model year 2001 improvements.

To complete our analysis of crash prevention, Figure 7-3 presents the numbers of vehicles involved in crashes under each Scenario,[130] while Figures 7-4 and 7-5 present the avoided and avoidable crashes in each model year group. For instance there were 2,610,000 cars in frontal crashes between January 1, 2008, and December 31, 2008. Had we replaced all model year 2006 and newer cars with model year 2005 cars, there would have been 2,634,000 cars in frontal crashes, an increase of 24,000 car involvements. That is, controlling for vehicle type, crash type, crash year, and vehicle age,[131] technologies that began to appear in appreciable numbers with model year 2006 cars are, in a sense, creditable with preventing 24,000 cars from getting in frontal crashes.

Had we replaced all model year pre-2005 cars with model year 2005 cars, there would have been 2,592,000 cars in frontal crashes, for a reduction of 18,000 car involvements. That is, to the extent that our models accurately predict crash rates, model year 2005 technologies could have prevented about 18,000 frontal car involvements in 2008.

---

[130] We have baselined Figure 7-3 to the actual vehicle crashes that occurred in 2008. That is, Figure 7-3 depicts the terms *VCActual*(*v*, *c*, 2008) + *VCAvoid* (*m*, *v*, *c*, 2008) and *VCActual*(*v*, *c*, 2008) – *VCAvoidable*(*m*, *v*, *c*, 2008), where *VCActual*(*v*, *c*, 2008) denotes the actual vehicle crashes in 2008 for vehicle type *v* and crash type *c*.

[131] As noted in chapter 5, a better crash avoidance model would control for driver impairment and driver experience, but we are not aware of data that would allow us to fit such a model.

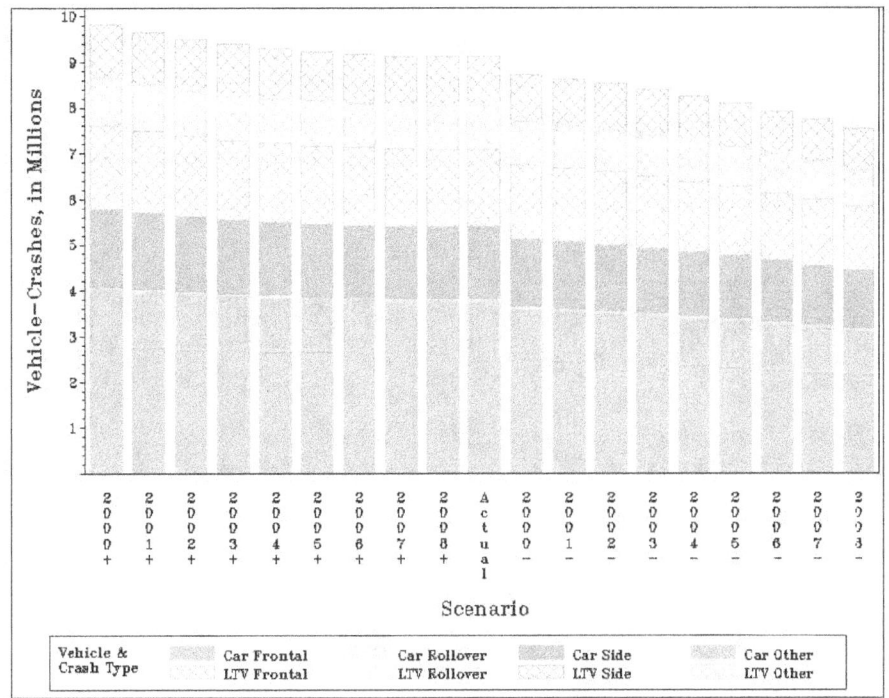

Figure 7-3: Vehicle Crashes in 2008, by Scenario, Crash Type, and Vehicle Type

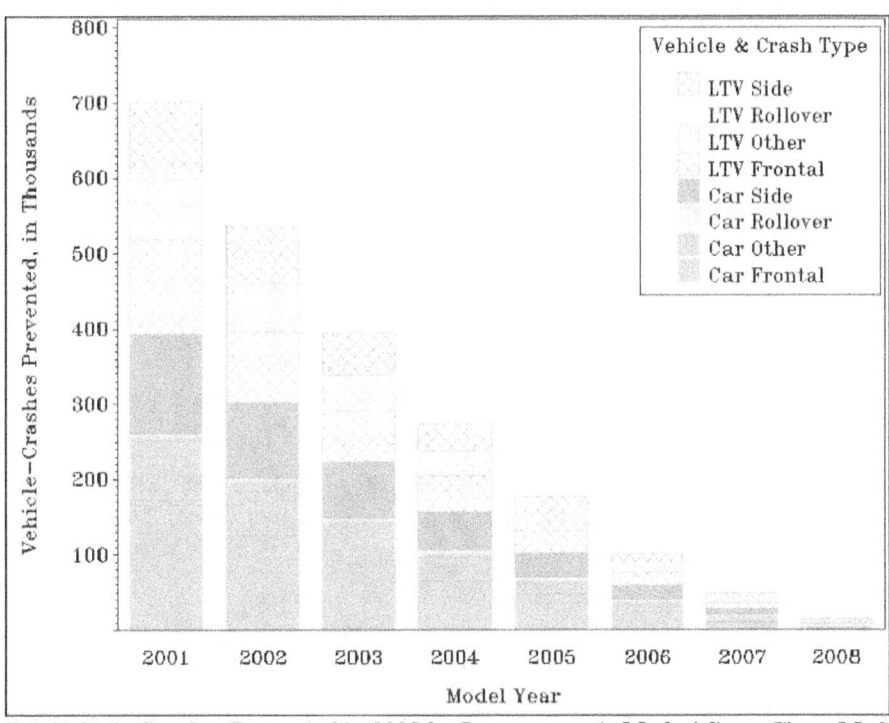

Figure 7-4: Vehicle Crashes Prevented in 2008 by Improvements Made After a Given Model Year

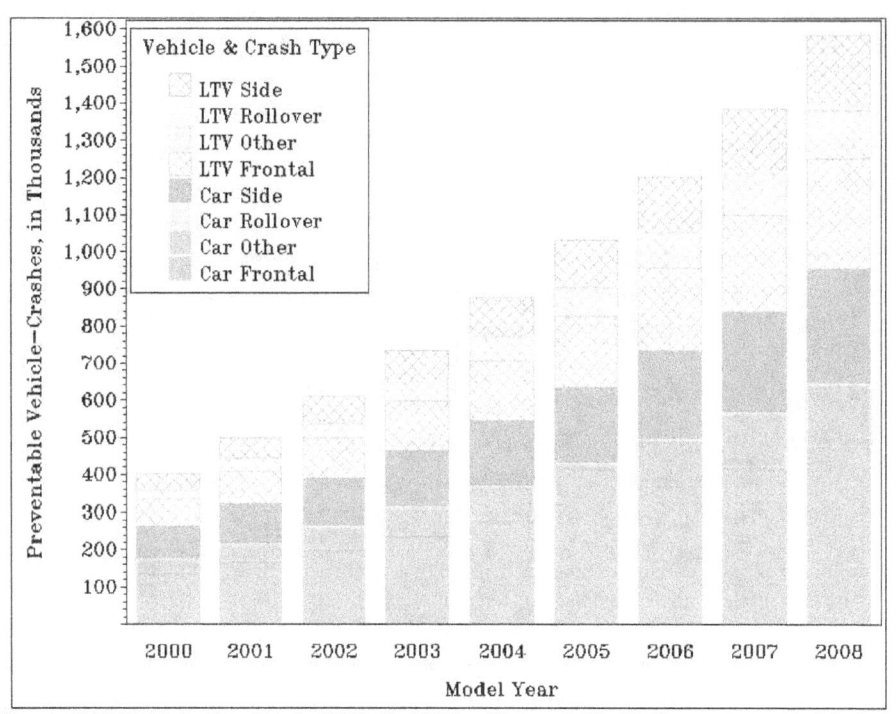

Figure 7-5: Vehicle Crashes in 2008 that Were Preventable by Improvements Made in a Given Model Year, Compared to the Actual Vehicles Driven

7.2 Injuries Mitigated and Mitigatable

Again, fix a calendar year y between 2000 and 2008, inclusive. We estimate the reduction $OccReduceScen(m, v, c, y, z)$ in the number of occupants injured at injury level z in vehicles of type v in crashes of type c that would result in year y under Scenario $m+$ as follows.

$OccReduceScen(m, v, c, y, z) =$

$$\sum_{k=m}^{y+1} \sum_{d,r,e,g} \left(InjDist_{dreg}(m, c, v, y, z) - InjDist_{dreg}(k, v, c, y, z) \right) \frac{VCAvoidScen(m, v, c, y, k)}{VCActual(v, c, y, k)} Occs_{dreg}(v, c, y, k, z)$$

where the component terms are as follows:

- $InjDist_{dreg}(m, c, v, y, z)$ denotes our model's predicted probability that the most severe injury sustained by an occupant of age group e and gender g with restraint use r in a crash of type c in a model year m vehicle of type v driven by a driver in alcohol category[132] d is of injury level z;[133]

- $VCActual(v, c, y, k)$ denotes the average of the nonmissing values among the five hotdeck estimates $VehCrashes_{ycvki}$ from Chapter 4; that is, $VCActual(v, c, y, k)$ estimates the number of model year k vehicles of type v in crashes of type c in calendar year y;

- $Occs_{dreg}(v, c, y, k, z)$ denotes the average of the nonmissing values among the five hotdeck estimates $Occs_{ycvkoi}$ from Chapter 4; that is, $Occs_{dreg}(v, c, y, k, z)$ estimates the number of occupants in age category e and gender g with restraint use r in model year k vehicles of type v driven by drivers in alcohol category d in crashes of type c in calendar year y; and

- $VCAvoidScen(m, v, c, y, k)$ denotes the reduction in the number of model year k vehicles of type v in crashes of type c in calendar year y under Scenario $m-$, as estimated from the techniques of the previous section, namely[134],

$$VCAvoidScen(m, v, c, y, k) = (CR(y + m - k, c, v, y - k) - CR(y, c, v, y - k))\, VehMiles_{yvk}$$

That is, we hypothetically resituate occupants of post model year m cars and LTVs into model year m ones, and estimate the increased number of resulting crashes using our earlier techniques, under the necessary presumption that the vehicles in the hypothetical new crashes have the same occupancy (occupants per vehicle) as those that actually crashed. We then compare the injury distributions of the replacement and replaced vehicles to tabulate the additional occupants injured at the given level in either an actual or a hypothetical crash. The term $OccReduceScen(m, v, c, y, z)$ estimates the number of occupants that weren't injured at level z in calendar year y because of collective improvements in the model year $m+1$ through 2009 fleets. It includes both those who weren't injured at all because their crash was avoided, as well as those were injured to a lesser degree.

We can simplify the formula for $OccReduceScen(m, v, c, y, z)$ to

$$OccReduceScen(m, v, c, y, z) = \sum_{k=m}^{y+1} \frac{VCAvoidScen(m, v, c, y, k)}{VCActual(v, c, y, k)} Occs(m, v, c, y, k, z)$$

where

$$Occs(m, v, c, y, k, z) = \sum_{d,r,e,g} \left(InjDist_{dreg}(m, c, v, y, z) - InjDist_{dreg}(k, v, c, y, z) \right) Occs_{dreg}(v, c, y, k, z)$$

As with crash reduction, we isolate the portion attributable to the individual fleets by taking successive differences:

$OccReduce(m, v, c, y, z) = OccReduceScen(m, v, c, y, z) - OccReduceScen(m-1, v, c, y, z)$

---

[132] Our alcohol categories are: alcohol involved, no alcohol involved, and no driver (i.e., the vehicle had no driver).
[133] For an occupant sustaining multiple injuries, z is the level of the most severe injury sustained.
[134] We are purposefully being overly broad in our notation for increased readability. Recall that side impacts are treated as one crash type in crash avoidance, but two (near side, far side) in crashworthiness. Thus when applying $VCAvoidScen(m,v,c,y,k)$ for c = near side (or far side) crashes, one should take the vehicle crash reduction for side impacts. As a result, our injury mitigation figures, in a sense, rest on the assumption that near and far side crashes are prevented in equal measure.

The term $OccReduce(m, v, c, y, z)$ estimates the number of occupants not injured at level z in year y because of improvements in the model year m fleet. It reflects injuries that didn't happen because their crashes didn't happen and injuries that were mitigated by improved crashworthiness. Thus, $OccReduce(m, v, c, y, z)$ is a measure of the combined improvements to crash avoidance and crashworthiness to the model year m fleet, presenting the net effect of the improvements for level z injuries.

We similarly estimate the expected increase $OccReducible(m, v, c, y, z)$ in the number of occupants injured at injury level z in vehicles of type v in crashes of type c that would result under Scenario $m-$ in year y as:

$$OccReducibleScen(m, v, c, y, z)$$
$$= \sum_{k=1974}^{m-1} \sum_{d,r,e,g} \left(InjDist_{dreg}(k, c, v, y, z) - InjDist_{dreg}(m, v, c, y, z) \right) \frac{VCAvoidableScen(m, v, c, y, k)}{VCActual(v, c, y, k)} Occs_{dreg}(v, c, y, k, z)$$
$$= -\sum_{k=m}^{y+1} \frac{VCAvoidableScen(m, v, c, y, k)}{VCActual(v, c, y, k)} Occs(m, v, c, y, k, z)$$

$OccReducibleScen(m, v, c, y, z)$ estimates the number of occupants whose level z injuries didn't have to happen because of the extent to which the model year m fleet is safer than those from model years $1974, \ldots, m-1$. We isolate the model year m contribution as:

$$OccReducible(m, v, c, y, z) = OccReducibleScen(m, v, c, y, z) - OccReducibleScen(m-1, v, c, y, z)$$

The term $OccReducible(m, v, c, y, z)$ estimates the number of occupants whose level z injuries didn't have to happen because of crash avoidance and crashworthiness improvements in the model year m fleet over the previous year's fleet.

Calculations on our datasets provide the following estimates of injury mitigation for the 2008 crash year.

Impact of Vehicle Improvements on Injuries – the Big Picture

Figure 7-6 presents the total reductions in the occupants at the various injury (KABCO) levels in 2008 resulting from improvements to individual model year fleets. That is, this chart shows the estimates $\sum_v \sum_c \sum_z OccReduce(m, v, c, 2008, z)$ in blue and $\sum_v \sum_c \sum_z OccReducible(m, v, c, 2008, z)$ in orange. (In both cases, v runs through the two vehicle types (car, LTV), c through the five crash types (frontal, near side, far side, rollover, other), and z through the five KABCO levels (O, C, B, A, K).)

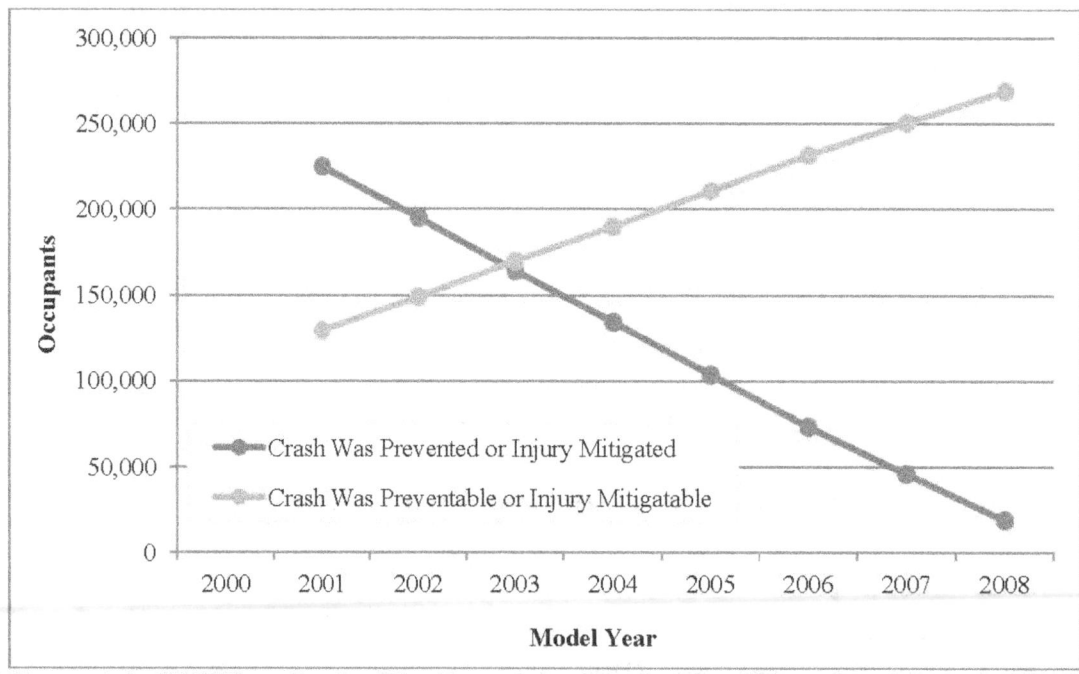

Figure 7-6: Occupants in 2008 Whose Crashes Were Prevented or Injuries Were Mitigated, and Those that Could Have Been Prevented or Mitigated, as a Result of Improvements to the Passenger Vehicle Fleet

As with crashes prevented, the appearance of decreased benefit with advancing fleet year is due to our measure ascribing benefit to the earliest model year possible, and to drivers of older vehicles (up to about 10 years in vehicle age) being more crash-prone than those of newer vehicles.

Injuries – Areas of Greater and Lesser Impact

As with crashes, the benefits of fleet improvements are disproportionately distributed, as are the potential benefits. The Chi-Square Test finds that the reduction in occupants from model year 2001 improvements is not proportionate among the five KABCO levels. But once again, this is not for the reasons one would (or might) expect.

It would seem natural to expect improved crashworthiness to lead to an *increase* in the number of occupants who walk away uninjured, and *decreased* numbers of injured occupants. This would particularly make sense if crashworthiness was improved without simultaneous improvements to crash avoidance. However crash avoidance *has* improved (quite a bit), and as a result, the number of uninjured occupants actually *declines* from fleet improvements. That is, the additional occupants walking away uninjured are more than compensated by the occupants whose crashes are avoided entirely, resulting in a net *decrease* in uninjured occupants. This is illustrated in the green bars for the "mitigated or prevented" categories in Figure 7-7.

In a sense, it makes more sense to apply the Chi-Square Tests to the KABCO levels that reflect injuries (C, B, A, and K). Even here, the mitigated and prevented injuries from the improvements to the model year 2001 fleet are disproportionately distributed ($p<0.0001$). The fleet improvements prevented or mitigated a larger share of the incapacitating injuries than one would expect (11% of the occupants with prevented and mitigated injuries were incapacitated, compared to 10% for actual injuries). In contrast, the model year 2008 fleet improvements largely benefited the "possibly injured" category (accounting for 67% of injuries prevented or mitigated by the 2008 fleet, compared to 64% for the 2001 fleet).

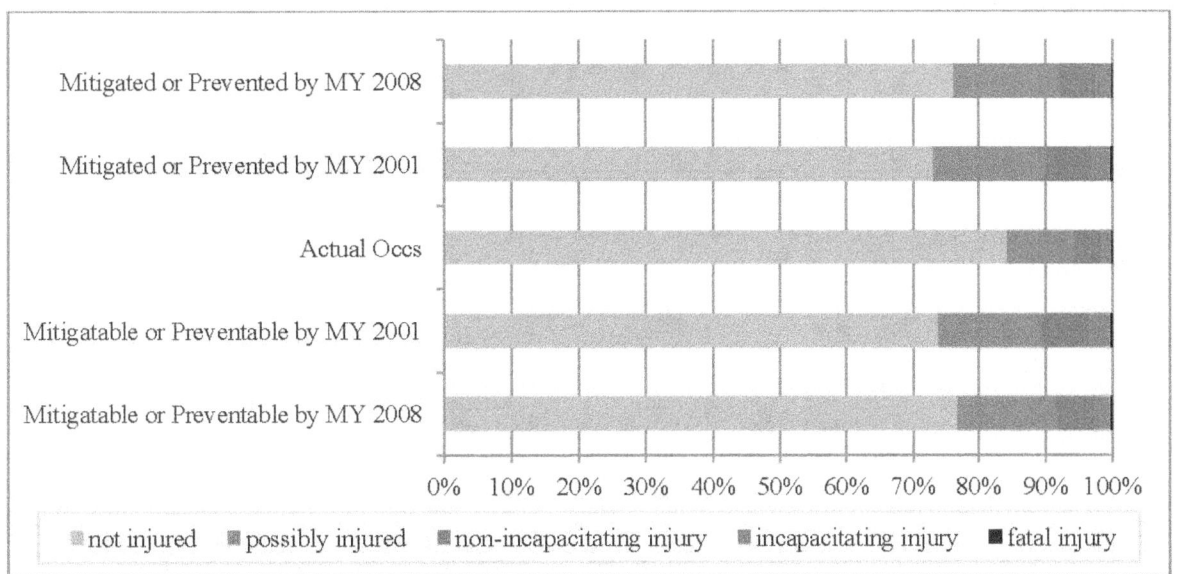

Figure 7-7: The Distribution of Occupants by Injury Level in 2008, and of Those with Prevented or Mitigated Injuries, or Preventable or Mitigatable Injuries, From Fleet Improvements

Figures 7-8 through 7-10 depict a similar analysis to inform the vehicle, crash, and occupant features in which most injury reduction has occurred. They show that the injury prevention and mitigation has occurred disproportionately more in side crashes, in LTVs, in vehicles with sober drivers and for restrained occupants, for 25- to 65-year-olds, and for women.

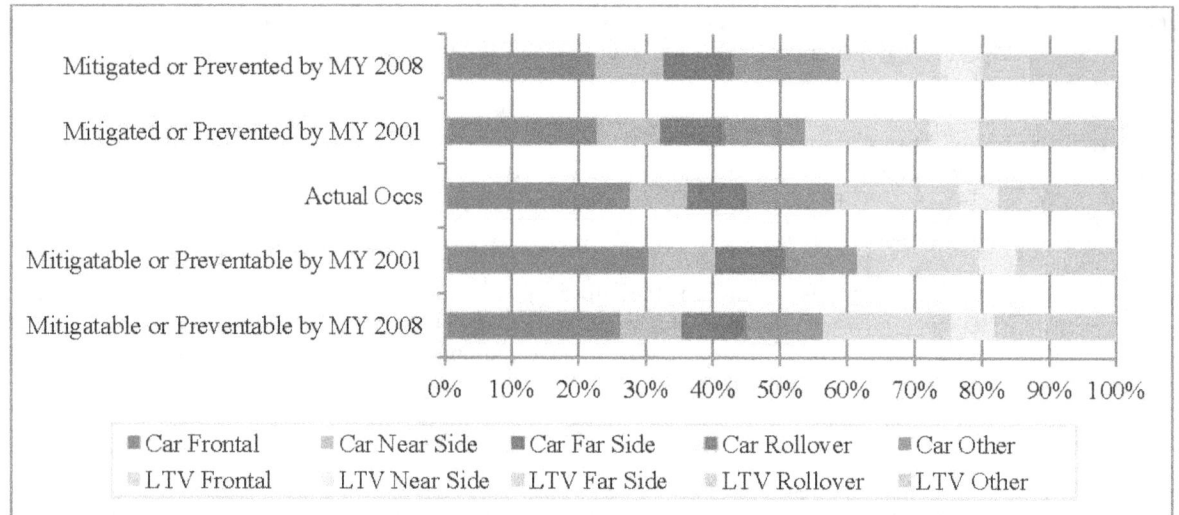

Figure 7-8: The Distribution of Occupants by Crash and Vehicle Type in 2008, and of Those Whose Injuries Were Prevented or Mitigated, or Preventable or Mitigatable, From Fleet Improvements

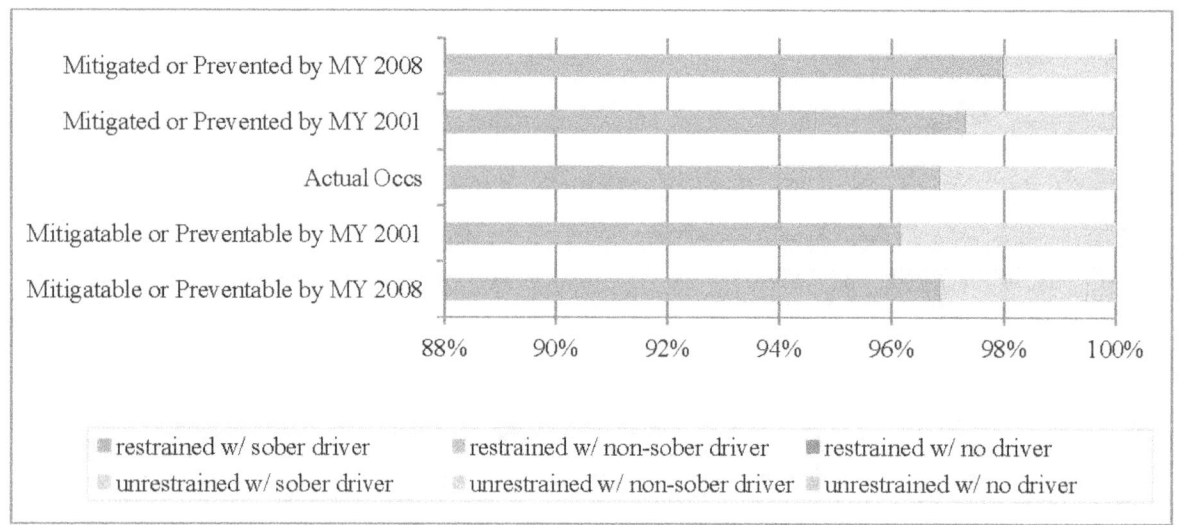

Figure 7-9: The Distribution of Occupants by Driver Sobriety and Occupant Restraint Use in 2008, and of Those Whose Injuries Were Prevented or Mitigated, or Preventable or Mitigatable, From Fleet Improvements

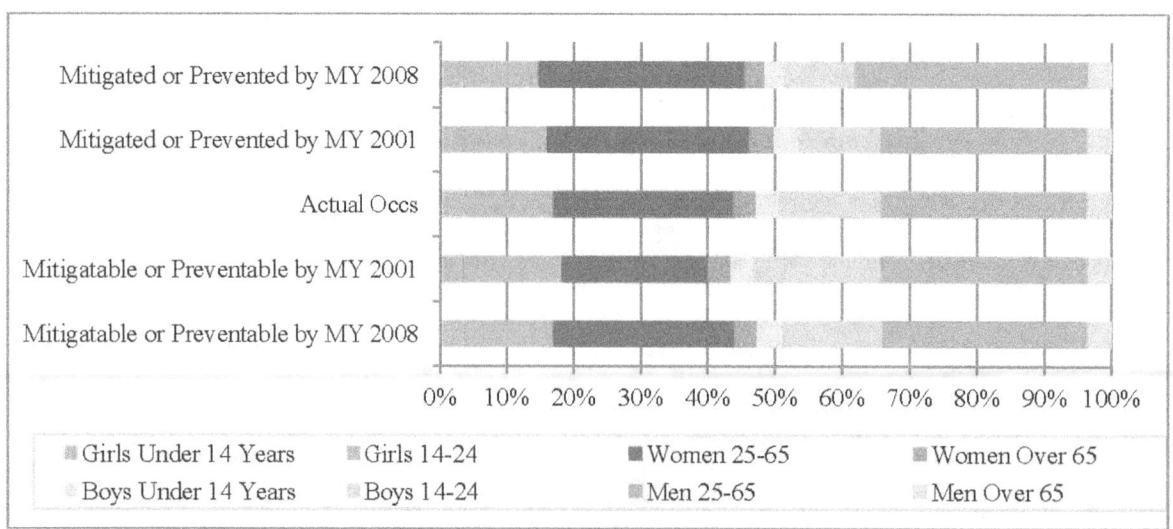

Figure 7-10: The Distribution of Occupants by Occupant Age and Gender, and of Those Whose Injuries Were Prevented or Mitigated, or Preventable or Mitigatable, From Fleet Improvements

Figure 7-11 presents the numbers of occupants injured at each level under each Scenario,[135] while Figures 7-12 and 7-13 present the occupants with prevented or mitigated, and preventable or mitigatable, injuries in each model year group. For instance there were 186,000 A-level injured car occupants in crashes between January 1, 2008, and December 31, 2008. Had we replaced all model year 2005 and newer cars with model year 2005 cars, there would have been 189,000 A-injured occupants, an increase of 3,000 occupants with incapacitating injuries. That is, controlling for crash type, vehicle type, driver alcohol, and the restraint use, age category, and gender of the occupant, crashworthiness technologies that began to appear in appreciable numbers with model year 2006 cars are, in a sense, creditable with mitigating the severity of 3,000 occupants that would otherwise have sustained A-level injuries.[136]

Had we replaced all model year pre-2005 cars with model year 2005 cars, there would have been 181,000 A-injured occupants, for a reduction of 6,000 occupants (difference due to rounding). That is, to the extent that our models take into account the major factors influencing crashworthiness, model year 2005 technologies could have mitigated the severity of about 6,000 A-injured occupants in 2008.

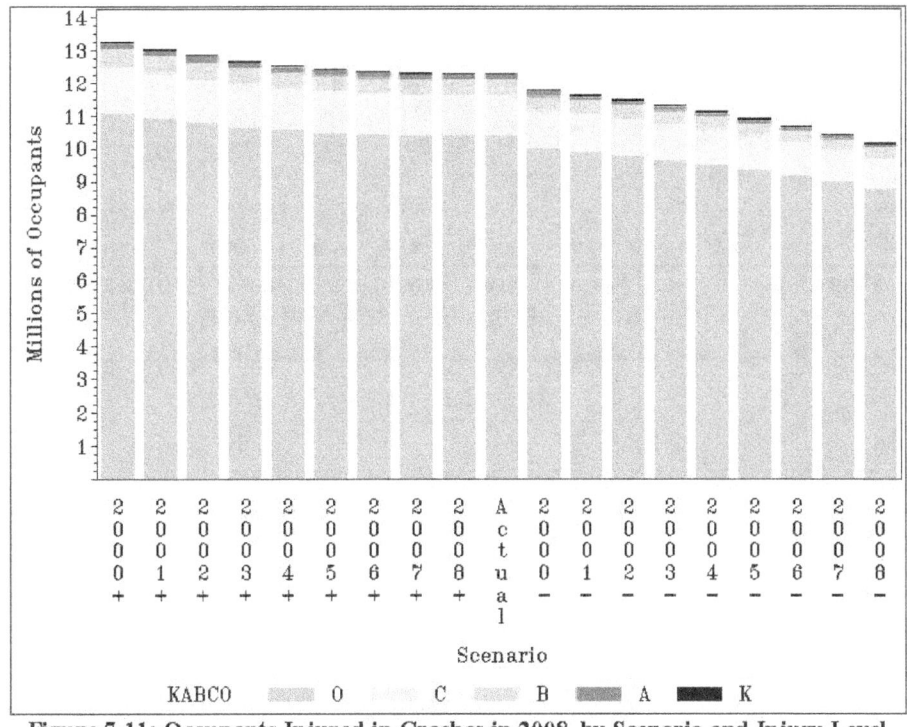

Figure 7-11: Occupants Injured in Crashes in 2008, by Scenario and Injury Level

---

[135] We have baselined Figure 7-11 to the actual occupants in crashes in 2008. That is, Figure 7-11 presents the terms $Occs_{2008}(z) + \sum_{v,c} OccReduce\ (m, v, c, 2008, z)$ and $Occs_{2008}(z) + \sum_{v,c} OccReducible\ (m, v, c, 2008, z)$, where $Occs_{2008}(z)$ denotes the estimated number of occupants injured at KABCO z in 2008.

[136] As noted in Chapter 6, a better crashworthiness model would utilize better measures of crash severity and other kinematics of the crash.

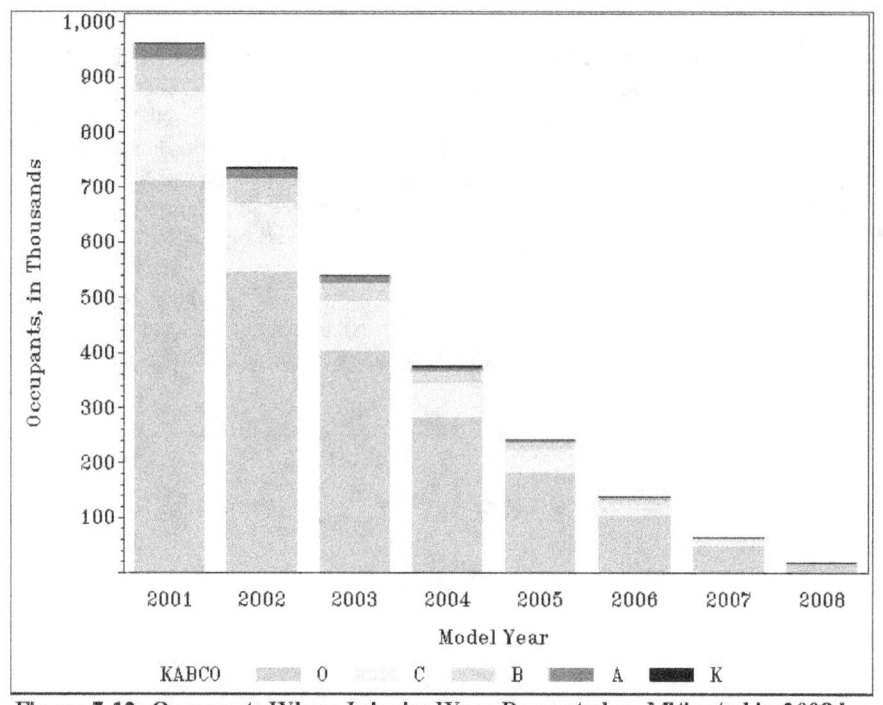

Figure 7-12: Occupants Whose Injuries Were Prevented or Mitigated in 2008 by Improvements Made After a Given Model Year

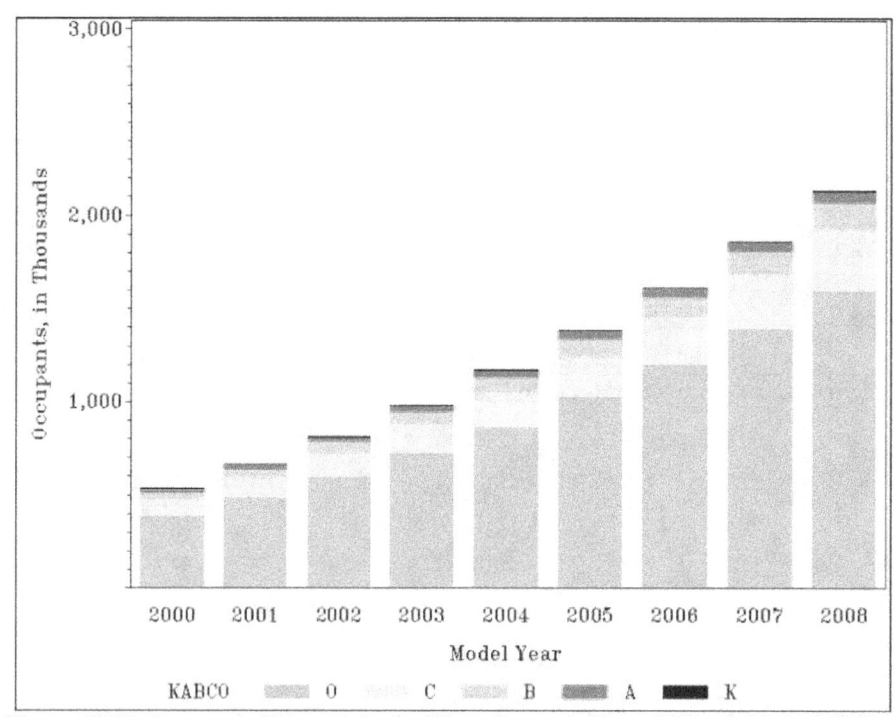

Figure 7-13: Occupants Whose Injuries Were Preventable or Mitigatable in 2008 by Improvements Made in a Given Model Year, Compared to the Actual Vehicles Driven

7.3 Lives Saved and Savable

Due to their particular interest, we isolate and explore further the results on lives saved and savable. Figure 7-14 presents the information from Figure 7-6, but limited to fatalities. Improvements to model year 2001 fleet saved 500 lives in 2008, and could have saved 400 more.

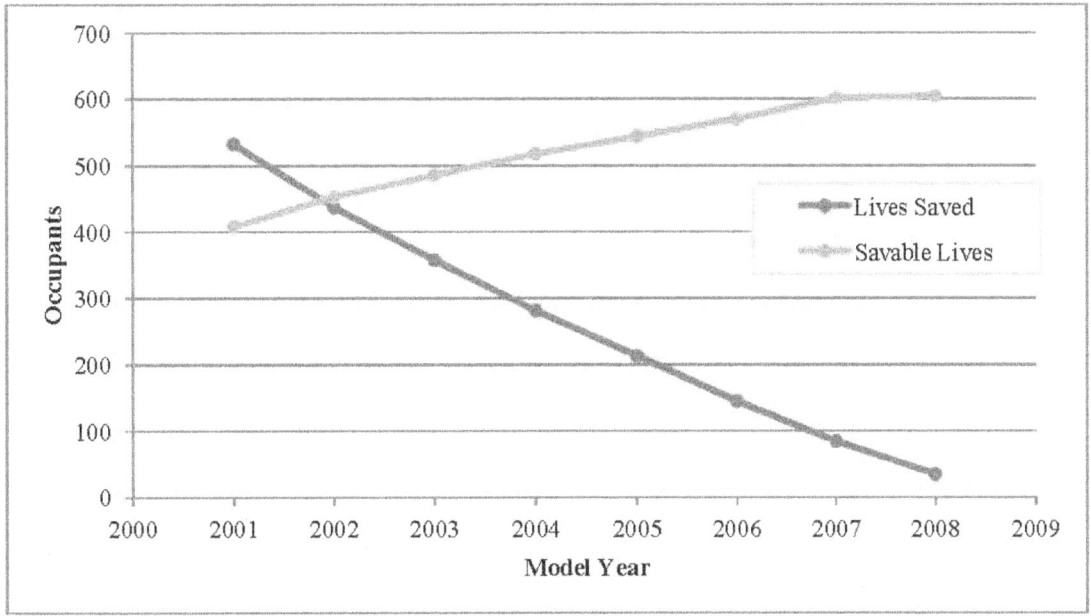

Figure 7-14: Saved and Savable Lives in 2008 From Improvements to the Passenger Vehicle Fleet

Figures 7-15 through 7-17 take the results of Figures 7-11 through 7-13 for K-level injuries and expand them by crash type. Improvements made after model year 2001 saved 2,000 lives in 2008, while an estimated additional 6,000 would have been saved had they been driving model year 2008 vehicles.

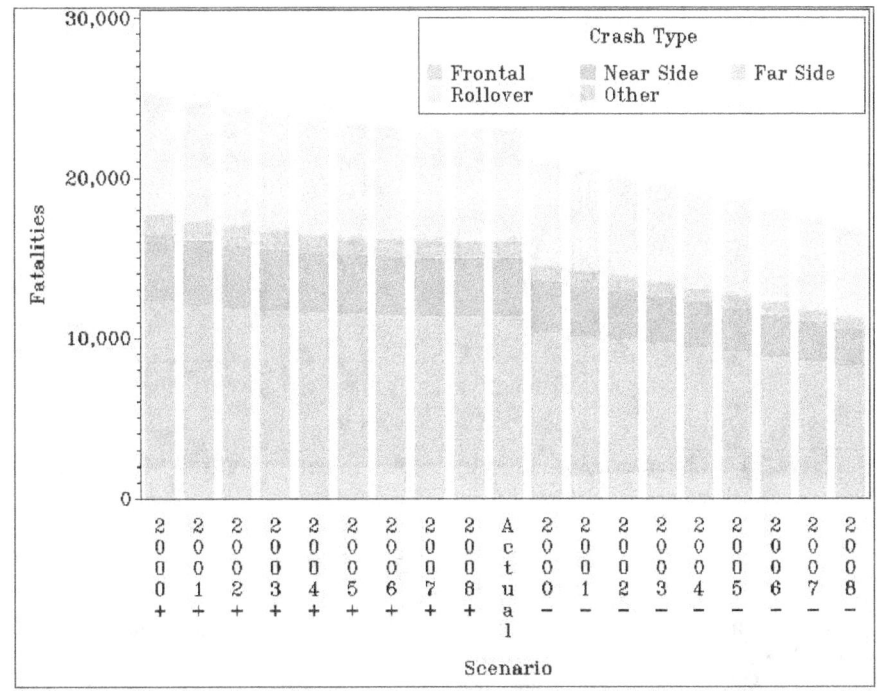

Figure 7-15: Passenger Vehicle Occupants Killed in Crashes in 2008, by Scenario and Crash Type

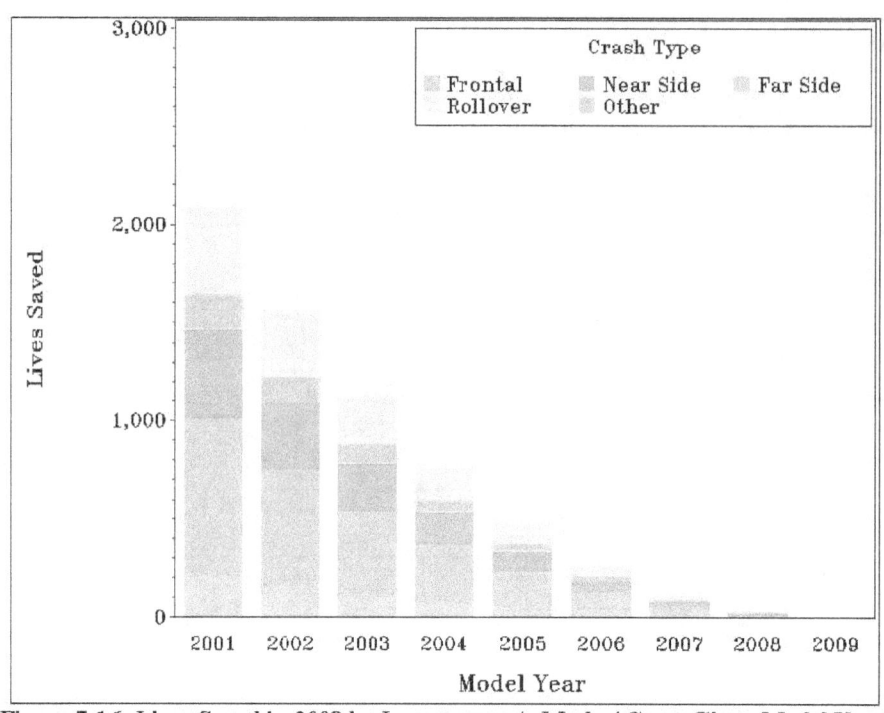

Figure 7-16: Lives Saved in 2008 by Improvements Made After a Given Model Year

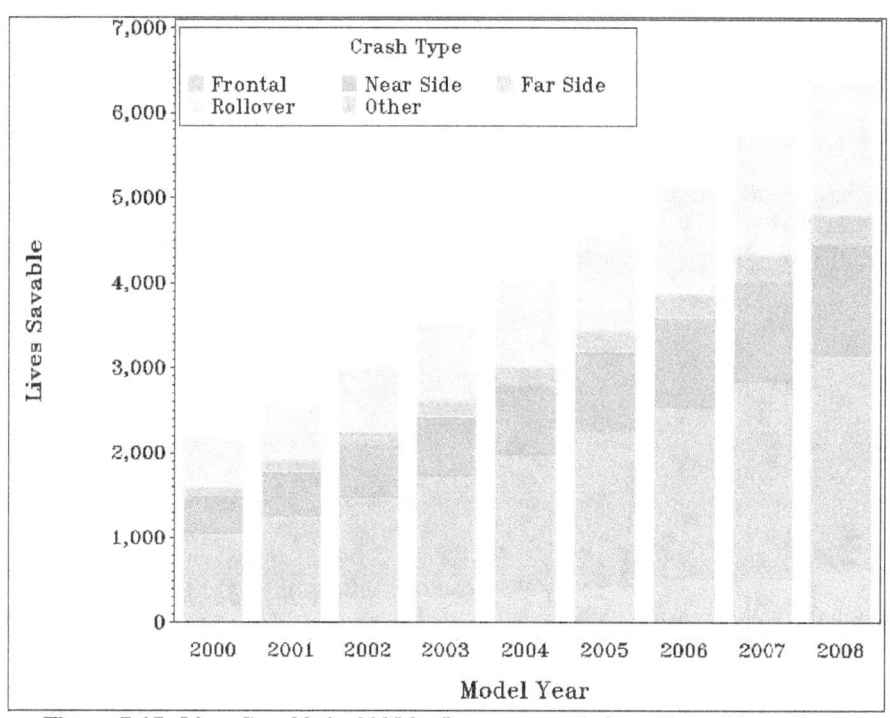

Figure 7-17: Lives Savable in 2008 by Improvements in a Given Model Year

7.4 The Benefits of Modeling in Assessing Fleet Improvements

In Chapter 4, we computed raw estimates of crash rates broken out by the factors in our crash avoidance model (crash type, crash year, vehicle type, and vehicle age). We also computed raw estimates of the likelihood of injury broken out by the factors in our crashworthiness model (crash type, vehicle type, model year, driver alcohol, occupant restraint use, occupant age group, and occupant gender). We could have used these raw estimates instead of the model estimates in estimating the avoided/avoidable crashes and mitigated/mitigatable injuries in this chapter, and they would have produced estimates of avoided/avoidable crashes and mitigated/mitigatable injuries that in a sense control for these several factors. The benefits of using the model estimates instead of the raw estimates are to provide estimates that are less susceptible to vagaries in the data and sampling error in small cells, and a particularly noteworthy benefit - allowing us to control for vehicle age in the estimates of avoidable crashes and mitigatable injuries.

For instance, consider the A-level injuries in frontal car crashes in the year 2008 that were prevented or mitigated by post-2004 model year improvements. Our estimate of this figure is *OccReduceScen(m, v, cy, y, z)*, with *m*=2005, *y*=2008, *c*=frontal, *v*=car, and *z*=A, and is computed as:

$$OccReduceScen(m, v, c, y, z) = \sum_{k=m}^{y+1} \frac{VCAvoidScen(m, v, c, y, k)}{VCActual(v, c, y, k)} Occs(m, v, c, y, k, z)$$

where

$$Occs(m, v, c, y, k, z) = \sum_{d,r,e,g} \left(InjDist_{dreg}(m, c, v, y, z) - InjDist_{dreg}(k, v, c, y, z) \right) Occs_{dreg}(v, c, y, k, z)$$

and

$$VCAvoidScen(m, v, c, y, k) = (CR(y + m - k, c, v, y - k) - CR(y, c, v, y - k)) VehMiles_{yvk}$$

recalling that *k* runs through the post-2004 model years, *d* through the driver alcohol statuses, and *r*, *e*, and *g* running through restraint use, age, and gender.

Consider the portion contributed by *k*=model year 2007, *d*=no driver, *r*=unrestrained, *e*=age 66+, and *g*=male. There might have been few, if any, unrestrained elderly males in driverless model year 2007 cars in frontal crashes in our crash database, and any cases that did occur might have a large weight in GES (e.g., if they occurred in a police jurisdiction with a small selection probability). Consequently the raw estimate of the injury likelihood for unrestrained elderly males in driverless model year 2007 cars in frontal crashes could well have a large sampling error. The raw estimate of the crash rate for model year 2007 cars in frontal crashes in 2008 could be unreliable if there are by chance an unusually high or low number of frontal crashes of model year 2007 cars in 2008, or crashes with unusually high or low GES sampling weights. Had we used raw estimates instead of model estimates for the terms *InjDist$_{dreg}$(i, c, v, y, ,z)* and/or *CR(y+m–i, c, v, y-i)*, for *i*=2007 and 2005, we could thus yield at an unduly high or low estimate of the term:

$$\left(InjDist_{dreg}(m, c, v, y, z) - InjDist_{dreg}(k, c, v, y, z) \right) \frac{VCAvoidScen(m, v, c, y, k)}{VCActual(v, c, y, k)} Occs(m, v, c, y, k, z)$$

This is just one term of *OccReduceScen(m, v, c, y, z)*, and one would hope any such undue over- and under-estimates would wash out in the summation. However they might not, or one might want to narrow the estimate of prevented and mitigated *A*-level injuries to e.g., those sustained by unrestrained elderly males in driverless model year 2005 cars in frontal crashes in 2008 (or some other example), where the chance of the small cell over- and under-estimations are less likely to wash out. Thus, in using our models to estimate avoided/avoidable crashes and mitigated/mitigatable injuries, we have reduced the susceptibility of our estimates to small cell sampling variation and anomalies in our data sources.

However, perhaps the greatest benefit of using model estimates is the ability to control for vehicle age in estimating avoidable crashes and mitigatable injuries. We saw in Chapter 5 that vehicle age has an appreciable effect on crash rates, perhaps because of the shifting cohort of drivers that a vehicle may undergo as it ages. Consider for example the frontal car crashes in 2008 that would have been prevented and the *A*-level injuries that would have been prevented or mitigated with model year 2005 technologies. One portion of these crashes occurred in model year 1998 cars, which at the time (i.e., in 2008) were 10 years old, when as we found in Chapter 5 are about at their peak. Comparatively, the replacement model year 2005 cars are at a relatively safe point in their lifetime, at three years of age. If we believe the vehicle age effect on crash rate is due to the people driving the vehicle, and not the vehicle itself, we should replace the 1998 car with a 2005 car as it would be driven at 10 years of age, yielding an appreciably higher crash rate than a 2005 car as it is driven at three years of age.

Our estimate of the model year 1998 frontal car crashes in 2008 that could have been avoided had these cars possessed model year 2005 technologies is:

$$VCAvoidable(m, v, c, y) = VCAvoidableScen(m, v, c, y) - VCAvoidableScen(m-1, v, c, y)$$

where

$$VCAvoidableScen(m, v, c, y) = \sum_{k=1974}^{m-1}(CR(y, c, v, y - k) - CR(y + m - k, c, v, y - k)) VehMiles_{yvk}$$

with *m*=2005, *y*=2008, *c*=frontal, and *v*=car. In the summand's term for model year *k*=1998, the crash rate for the model year 2005 replacement cars is *CR(y+m–k,c,v,y-k)* = *CR*(2015,frontal, car, 10), i.e., the crash rate that model year 2005 cars are predicted to have when they are 10 years old (in the year 2015). We have no raw estimate for this, and such a calculation is only possible with the model estimates. Likewise our mitigatable injuries estimate depends on *VCAvoidableScen(m, v, c, y, k)*, and so cannot be estimated from the raw estimates in a manner that controls for the age of the replacement vehicles.

If we give up trying to control for vehicle age and simply apply the raw estimates, one obtains the results in the left panels of Figures 7-18 through 7-23. These are the analog of Figures 7-3 through 7-5 and 7-11 through 7-13 that would be obtained without the use of the statistical models developed in this report. They appear to be somewhat nonsensical, with the negative results in Figures 7-19, 7-20, 7-22, and 7-23 indicating decreased vehicle safety. Although not provable, we feel that the data sufficiently strongly suggests a relationship between crash propensity and vehicle age (or driver cohort) that the lack of control for this key variable makes these results incorrect.

The rightmost panels of Figures 7-18 through 7-23 depict the results one would get if one applied our statistical models, but in the alternative manner we considered (and rejected) in Section 6.1. That is, these results reflect a view that the year of driving has a greater impact on crash propensity than the age of the vehicle.[137] As with using the raw estimates, the calculations under this alternative assumption indicate diminished vehicle safety and we feel the lack of control for vehicle age is the reason why. In a sense, the primary benefit of our approach is that it recognizes that in replacing a model year, say, 2009 car with a model year 2005 car, we are only replacing the car, and not its driver.

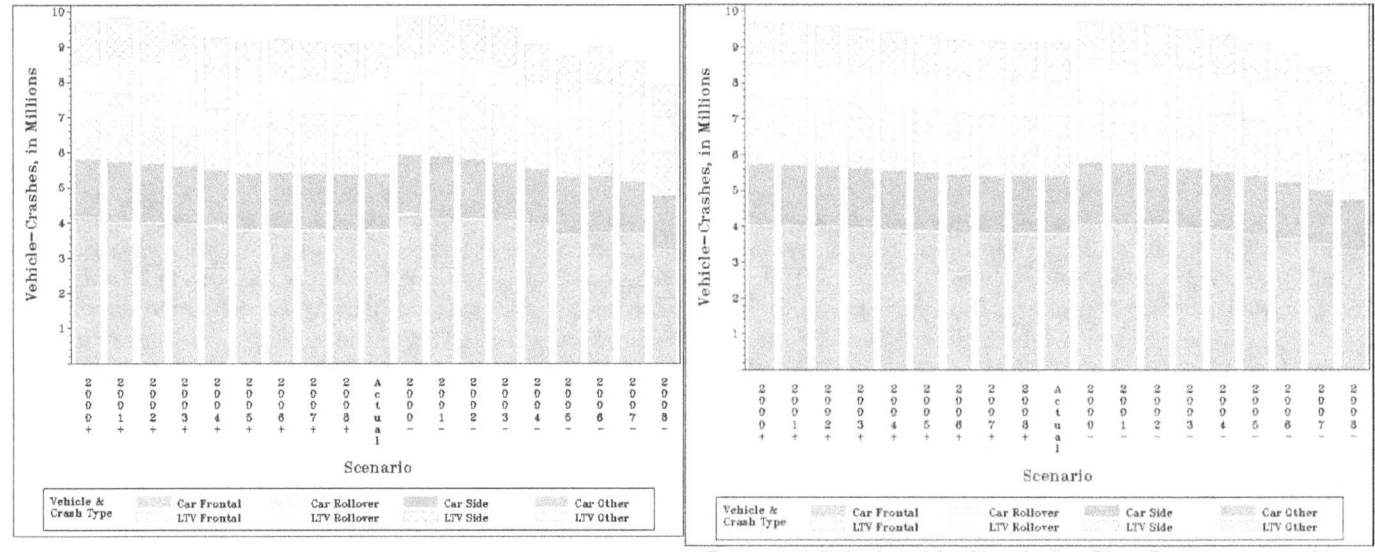

Estimates Calculated From Raw Data (No Modeling)
Estimates Calculated by Simulating Fleet Improvements by Reducing Vehicle Age in the Crash Avoidance Model

Figure 7-18: Vehicle Crashes in 2008, by Scenario, Crash Type, and Vehicle Type, Under Alternative Assumptions

---

[137] Recall that we unfortunately cannot control simultaneously for the year in which driving is conducted and the age of the vehicle when studying fleet improvements, since any two of calendar year, vehicle age, and model year determine the third.

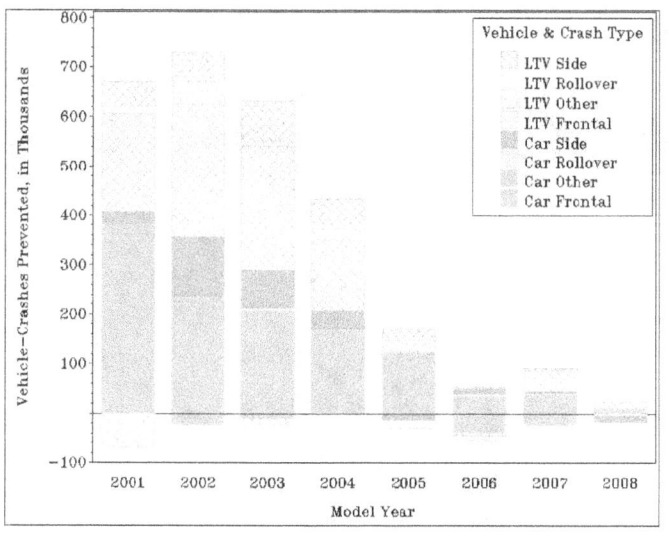 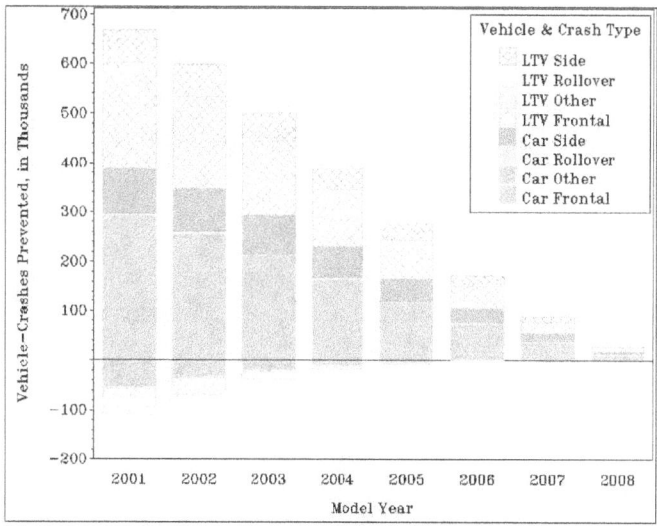

Estimates Calculated From Raw Data   Estimates Calculated by Simulating Fleet Improvements by Reducing Vehicle Age in the Crash Avoidance Model

Figure 7-19: Vehicle Crashes Prevented in 2008 by Improvements After a Given Model Year, Under Alternative Assumptions

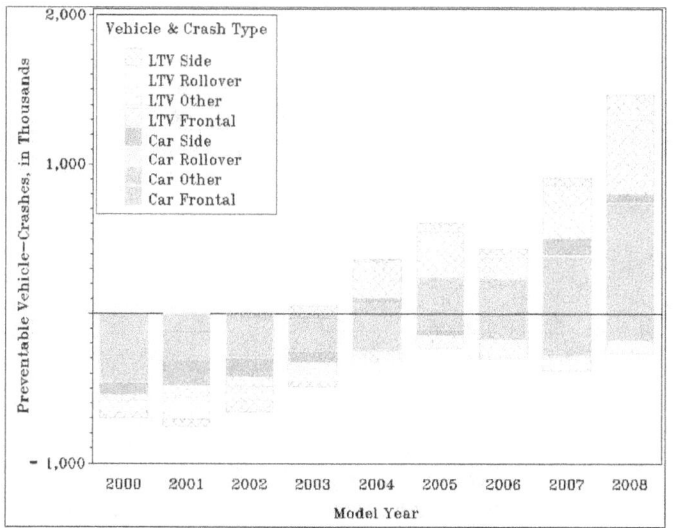 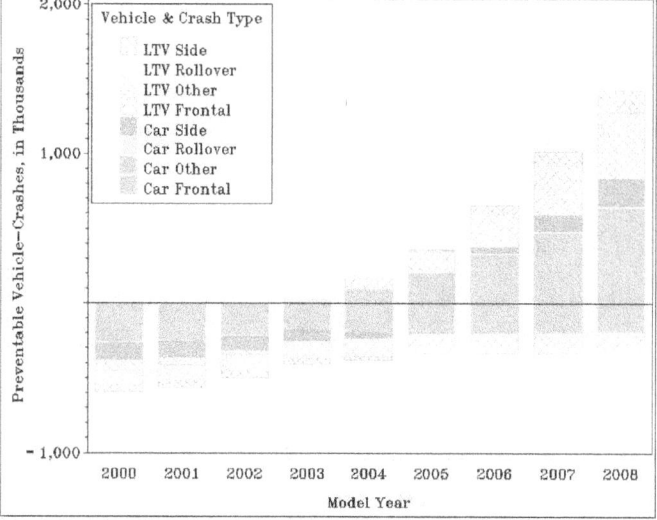

Estimates Calculated From Raw Data   Estimates Calculated by Simulating Fleet Improvements by Reducing Vehicle Age in the Crash Avoidance Model

Figure 7-20: Vehicle Crashes in 2008 That Were Preventable by Improvements in a Given Model Year, Under Alternative Assumptions

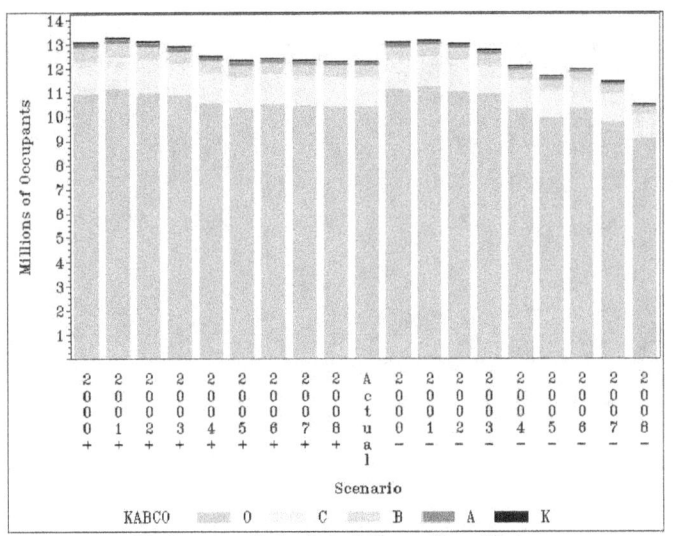
Estimates Calculated From Raw Data

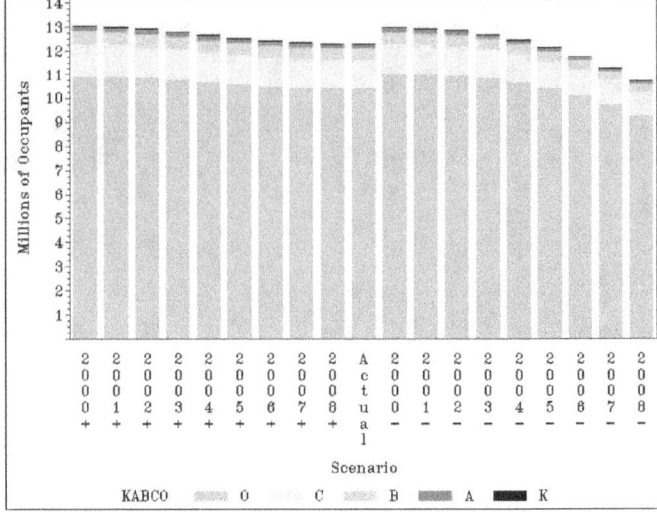
Estimates Calculated by Simulating Fleet Improvements by Reducing Vehicle Age in the Crash Avoidance Model

Figure 7-21: Occupants Injured in Crashes in 2008 by Scenario and Injury Level, Under Alternative Assumptions

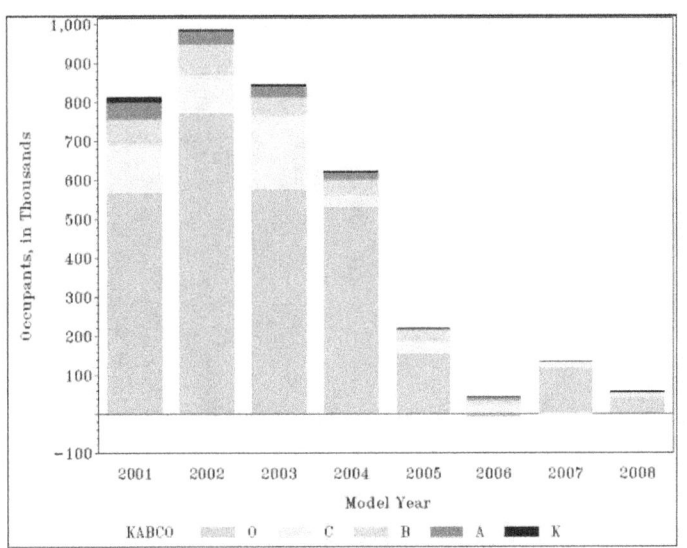
Estimates Calculated From Raw Data

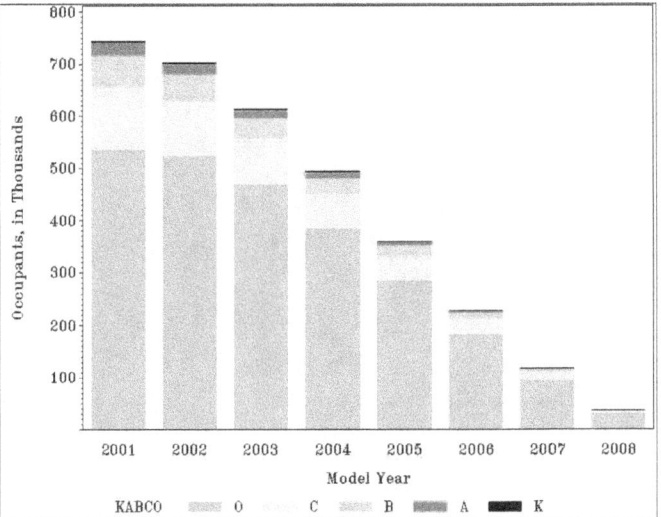
Estimates Calculated by Simulating Fleet Improvements by Reducing Vehicle Age in the Crash Avoidance Model

Figure 7-22: Occupants Whose Injuries Were Prevented or Mitigated in 2008 by Improvements Made After a Given Model Year, Under Alternative Assumptions

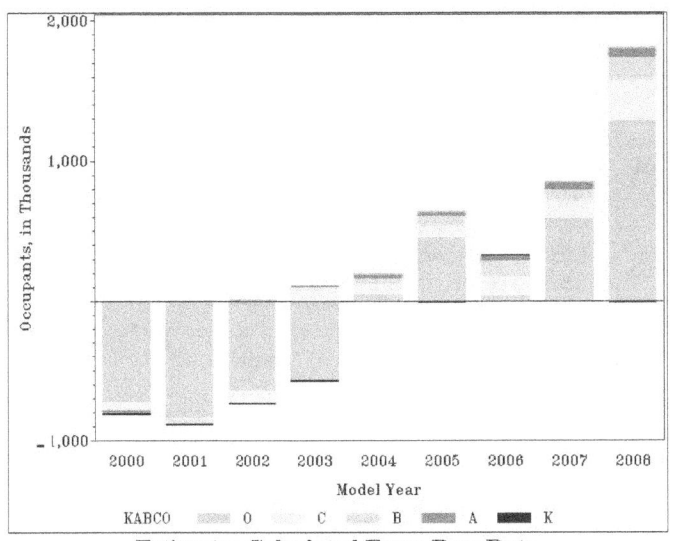 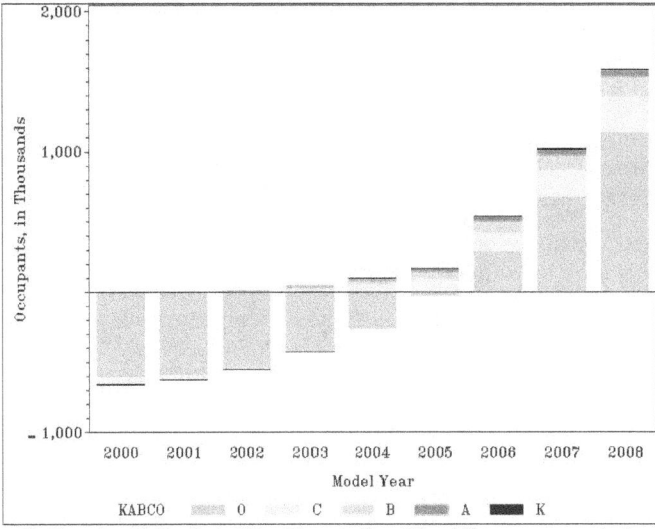

| Estimates Calculated From Raw Data | Estimates Calculated by Simulating Fleet Improvements by Reducing Vehicle Age in the Crash Avoidance Model |

Figure 7-23: Occupants Whose Injuries Were Preventable or Mitigatable in 2008 by Improvements Made in a Given Model Year, Under Alternative Assumptions

A Repeated Note on Attribution

As noted earlier, we do not attempt to identify the primary agents that drove the improvements seen in this report. The improvements to crash avoidance and crashworthiness, crashes prevented, lives saved, and injuries mitigated might be due to any of a host of technologies or design features. They might stem from Federal actions mandating a technology in the Federal Motor Vehicle Safety Standards, Federal requirements for optional equipment, the addition of a technology to the New Car Assessment Program as a recommended technology or through a star rating evaluating a level of performance, or the voluntary action of vehicle manufacturers to add technologies they are not required to add. It is even possible that some of the benefit derives from technologies or features that are not "new", but rather are technologies that have been around for a while and have gained market penetration. All we can say a priori about the source of the substantial benefits quantified in this report is that they stem from ways in which a new vehicle differs from an older one. The attribution of the benefits of safer vehicles to particular technologies and features is beyond the scope of this report.

8. References

Chang, D. (2009). *FARS Analytic Reference Guide 1975-2008, DOT HS 811 137*. Washington, DC: U.S. Department of Transportation, National Highway Traffic Safety Administration.

D16 Committee on Classification of Motor Vehicle Traffic Accidents. (2007). *American National Standard, Manual on Classification of Motor Vehicle Traffic Accidents, Seventh Edition, ANSI D-16.1– 2007*. Itasca, IL: National Safety Council.

Glassbrenner, D. (2011). An Analysis of Improvements to Vehicle Safety and Their Contribution to Recent Declines in Fatalities and Injury Rates, Paper No. 11-0228. *22nd Enhanced Safety of Vehicles Conference Proceedings*, (pp. 1-13). Washington, DC: U.S. Department of Transportation, National Highway Traffic Safety Administration.

Kahane, C. (2004). *Lives Saved by the Federal Motor Vehicle Safety Standards and Other Vehicle Safety Technologies, 1960-2002, DOT HS 809 833*. Washington, DC: U.S. Department of Transportation.

Lu, S. (2006). *Vehicle Survivability and Travel Mileage Schedules, DOT HS 809 952*. Washington, DC: U.S. Department of Transportation.

National Highway Traffic Safety Administration. (2008). *2008 FARS Coding and Validation Manual*. Washington, DC: U.S. Department of Transportation.

National Highway Traffic Safety Administration. (2008). *General Estimates System Coding And Editing Manual 2008*. Washington, DC: U.S. Department of Transportation .

National Highway Traffic Safety Administration. (2008). *National Automotive Sampling System (NASS) General Estimates System (GES) Analytical User's Manual 1988-2008*. Washington, DC: U.S. Department of Transportation.

National Highway Traffic Safety Administration. (2008). *Traffic Safety Facts: Speeding*. Washington, DC: U.S. Department of Transportation.

NHTSA's National Center for Statistics and Analysis. (2009). *2008 Traffic Safety Annual Assessment - Highlights, DOT HS 811 172*. Washington, DC: U.S. Department of Transportation.

NHTSA's National Center for Statistics and Analysis. (2001). *Traffic Safety Facts 2000, DOT HS 809 337*. Washington, DC: U.S. Department of Transportation.

NHTSA's National Center for Statistics and Analysis. (2009). *Traffic Safety Facts 2008, DOT HS 811 170*. Washington, DC: U.S. Department of Transportation.

NHTSA's National Center for Statistics and Analysis. (2010). *Traffic Safety Facts 2009, DOT HS 811 402*. Washington, DC: U.S. Department of Transportation.

NHTSA's National Center for Statistics and Analysis. (2008). *Traffic Safety Facts: Passenger Vehicles*. Washington, DC: U.S. Department of Transportation.

Poch, M., & Mannering, F. (1996). Negative Binomial Analysis of Intersection-Accident Frequencies. *Journal of Transportation Engineering*, 105-113.

Rudin, W. (1976). *Principles of Mathematical Analysis*. New York: McGraw-Hill.

Strashny, A. (2007). *An Analysis of Motor Vehicle Rollover Crashes and Injury Outcomes, DOT HS 810 741*. Washington, DC: U.S. Department of Transportation.

Summers, S. M., Hollowell, W. T., & Prasad, A. (2003). NHTSA's Research Program for Vehicle Compatibility, Paper No. 307. *Enhanced Safety of Vehicles Conference Proceedings* (pp. 1-9). Washington, DC: Enhanced Safety of Vehicles Conference .

Turner-Fairbank Highway Research Center. (2005). *Revised Monograph of Traffic Flow Theory: A State-of-the-Art Report*. Washington, DC: Transportation Research Board, National Academy of Sciences.

9. Appendix

9.1 Raw Crashworthiness Estimates

The figures in this section present raw estimates with linear fits, pooling the crash data from the 2000-2008 calendar years.[138]

Figure A-1: Log-Odds Crashworthiness for Belted 25-65 Year Old Car Occupants With Sober Drivers

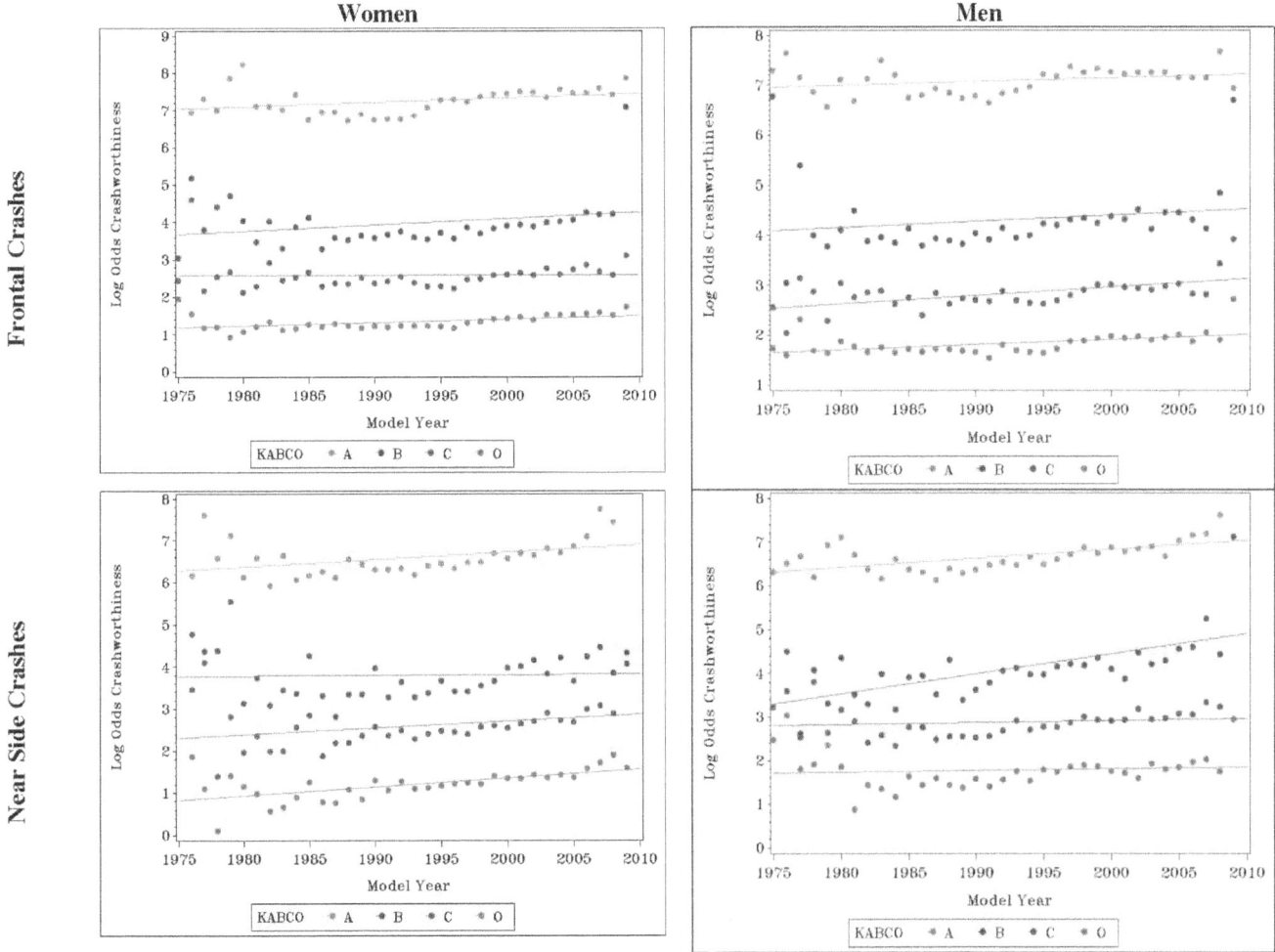

---

[138] Because many of the figures in this section break across pages, we depart from our convention and place figure titles above the figures.

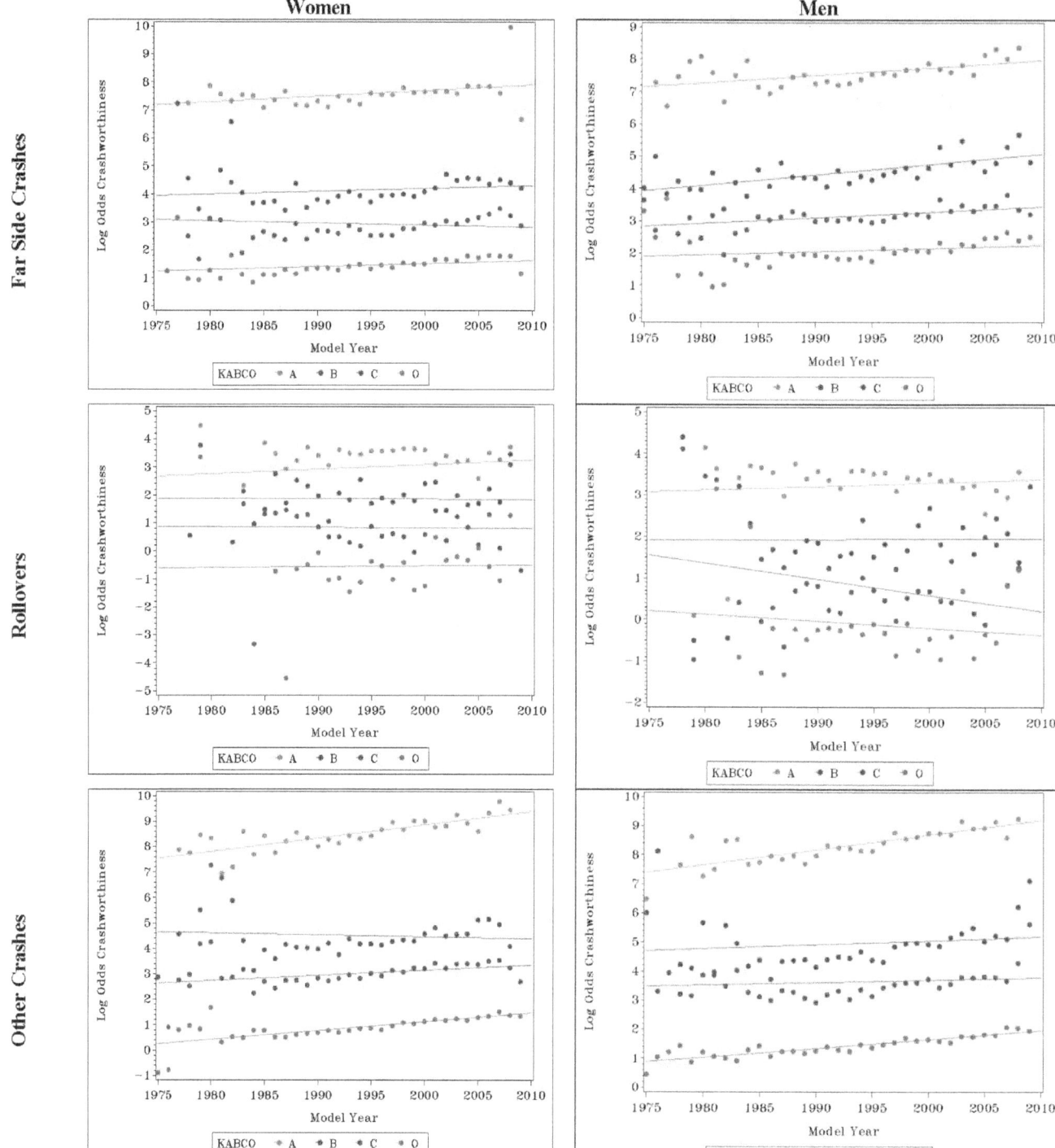

Figure A-2: Log-Odds Crashworthiness for Unbelted 25- to 65-Year-Old Car Occupants With Sober Drivers

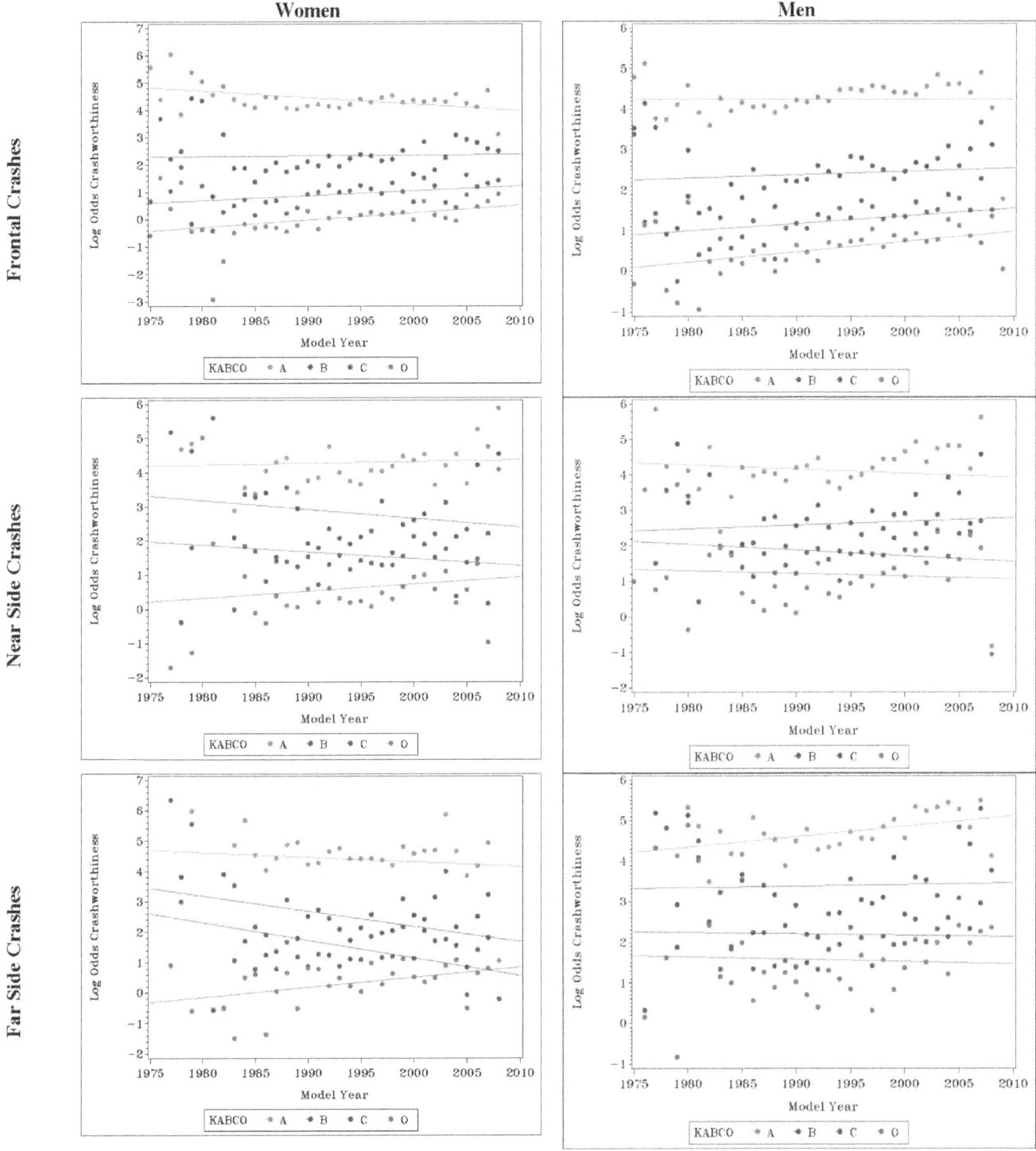

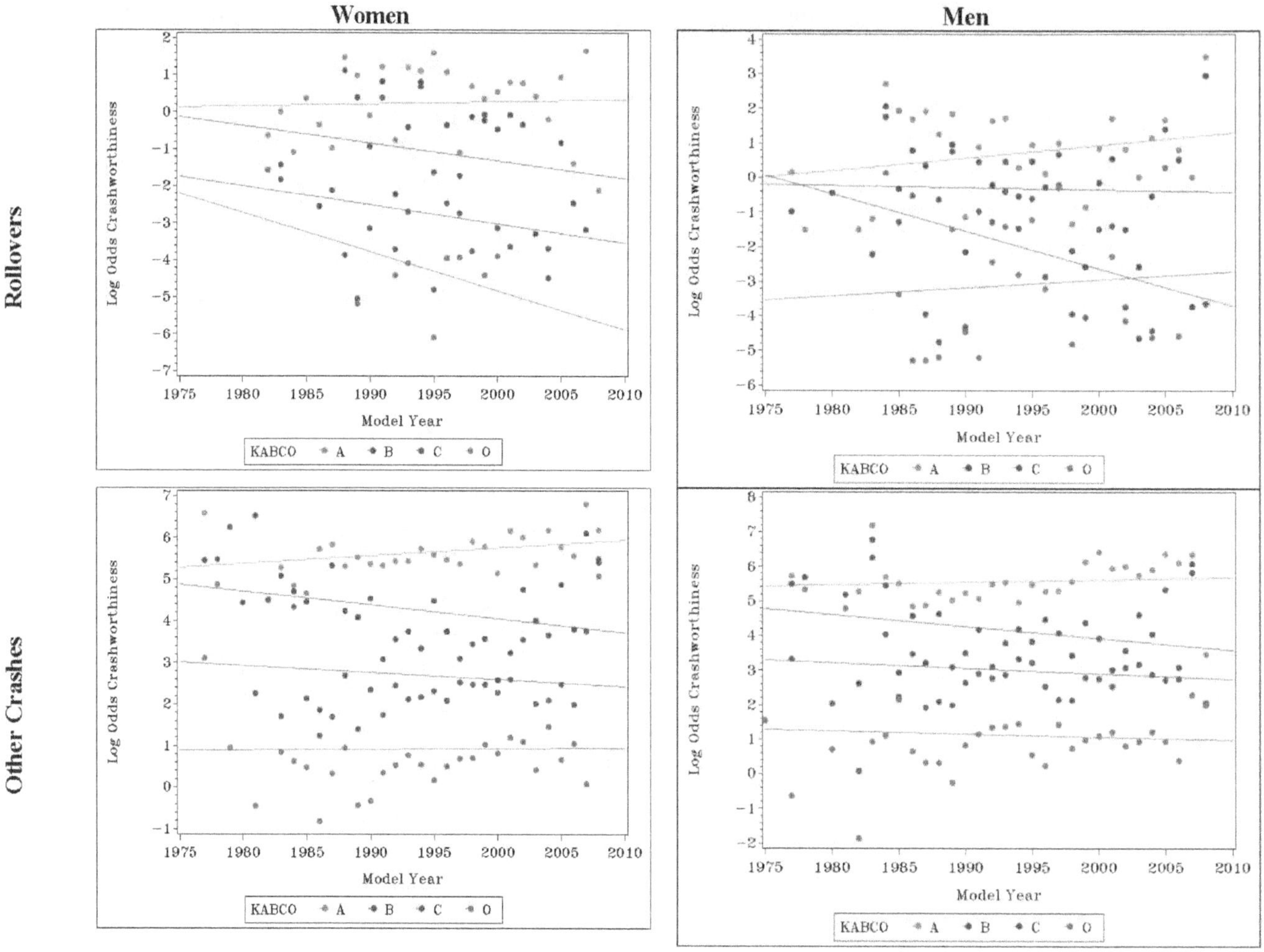

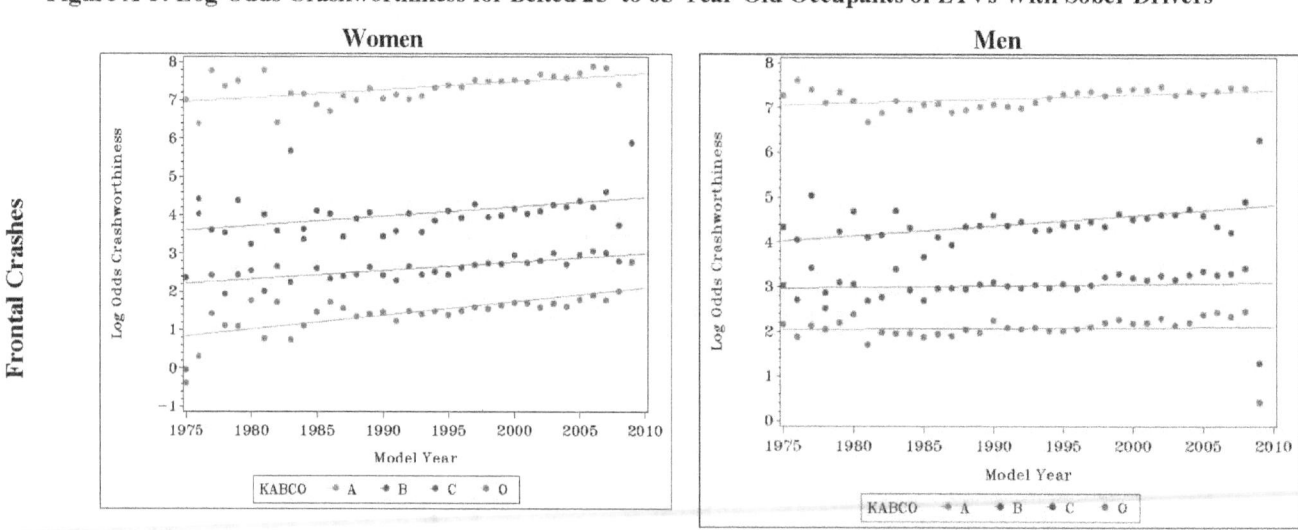

Figure A-3: Log-Odds Crashworthiness for Belted 25- to 65-Year-Old Occupants of LTVs With Sober Drivers

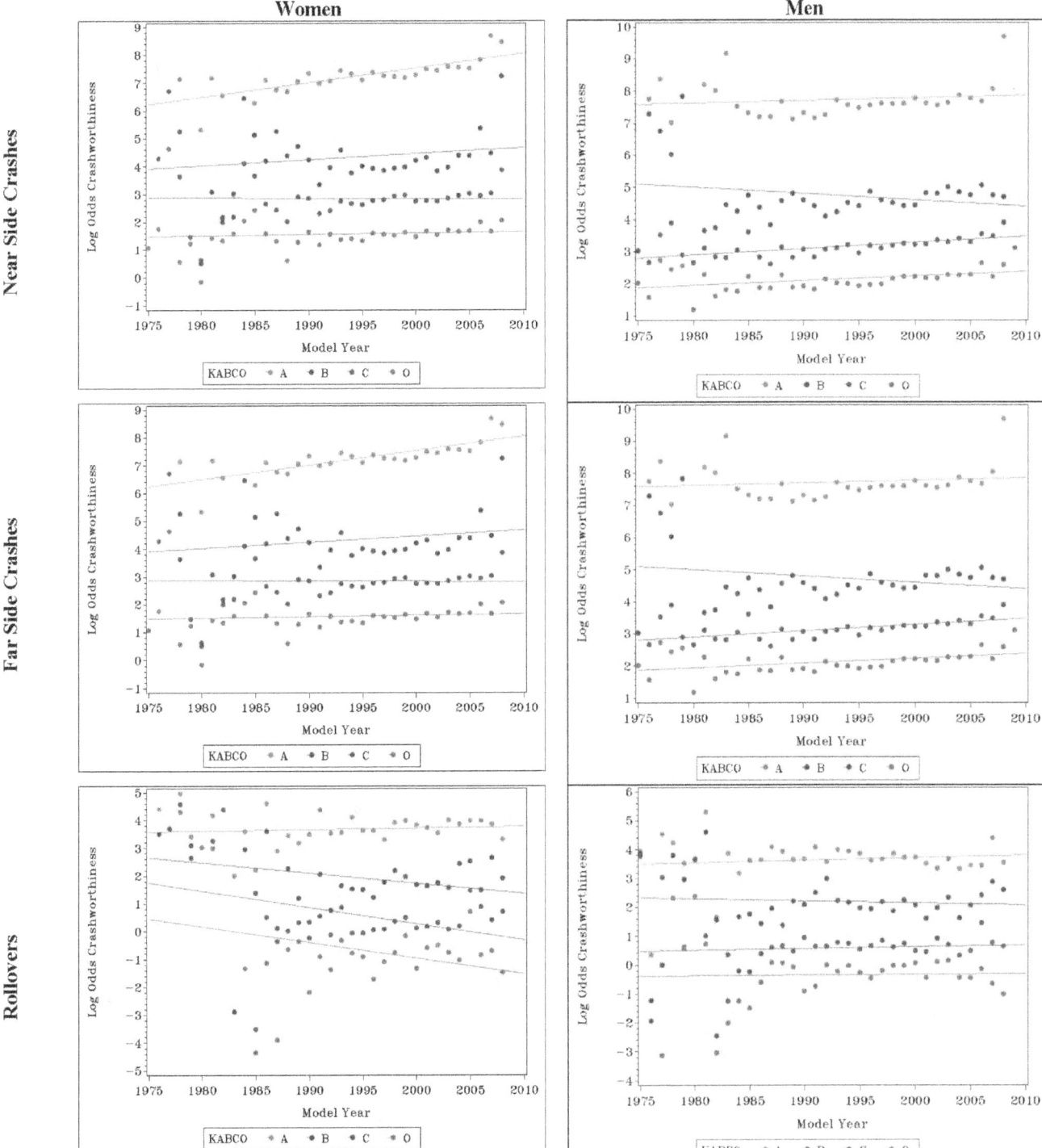

Other Crashes

Women | Men

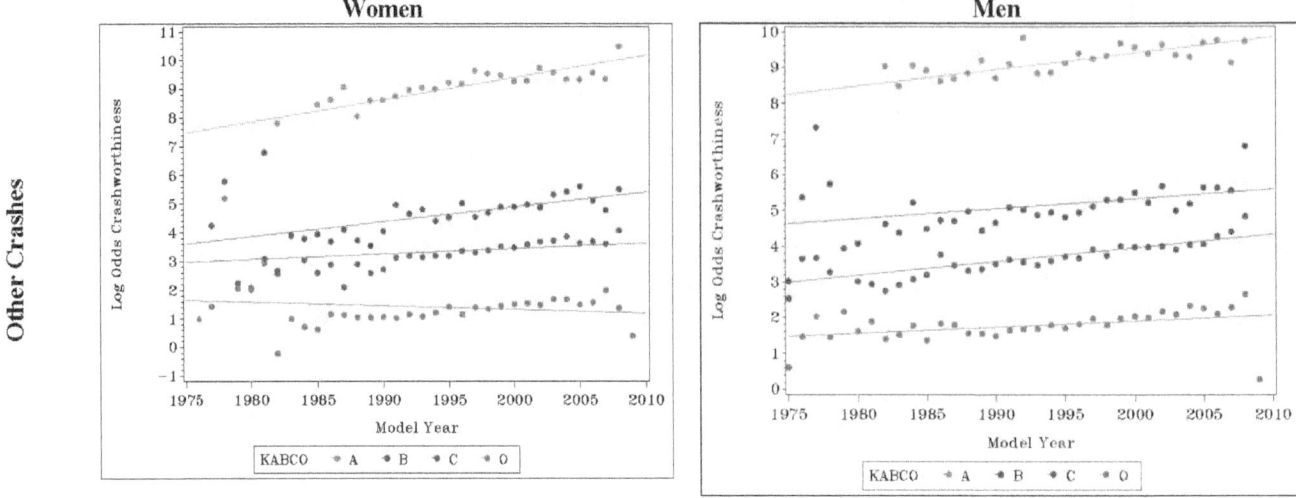

Figure A-4: Log-Odds Crashworthiness for Unbelted 25- to 65-Year-Old Occupants of LTVs With Sober Drivers

Women | Men

Frontal Crashes

Near Side Crashes

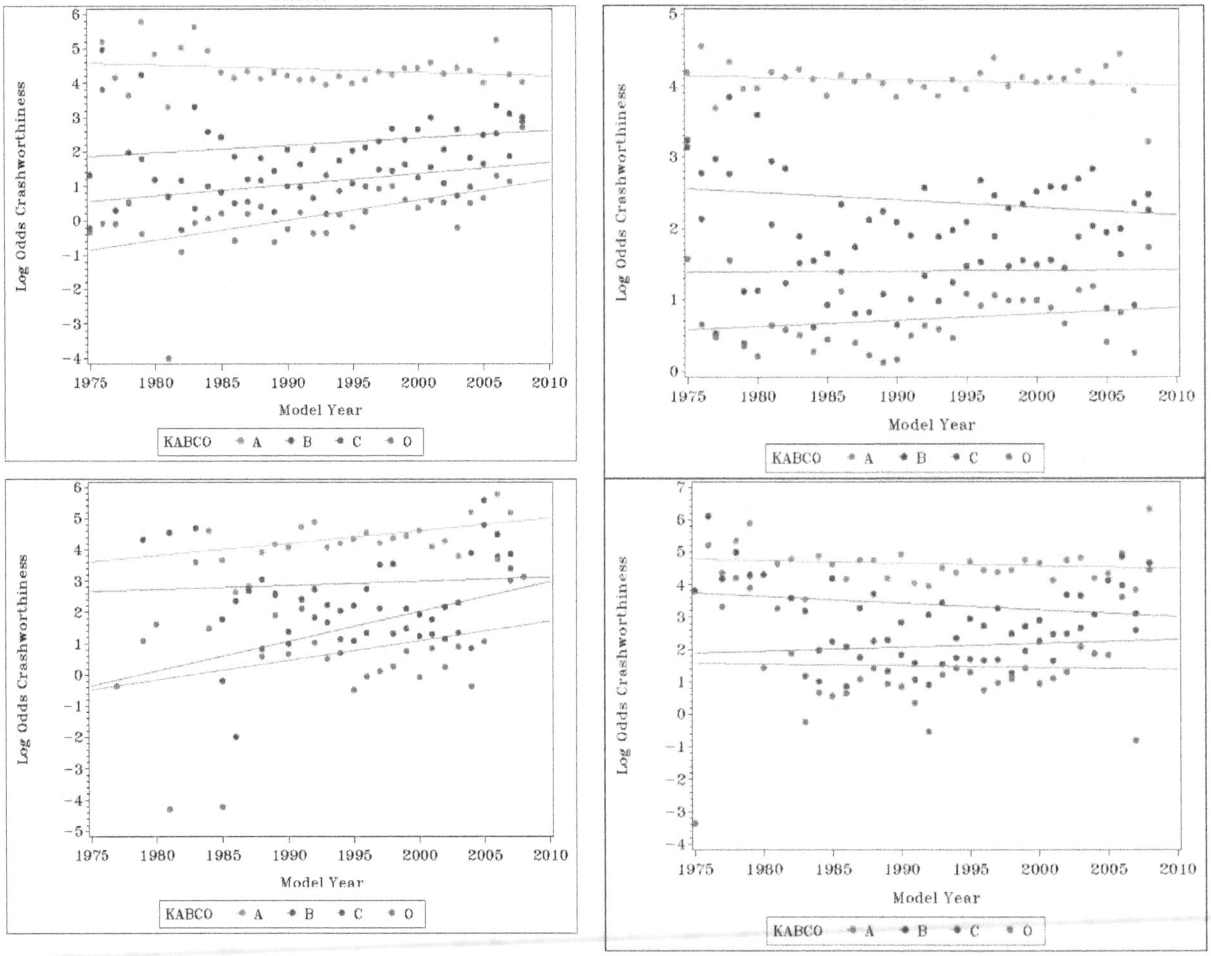

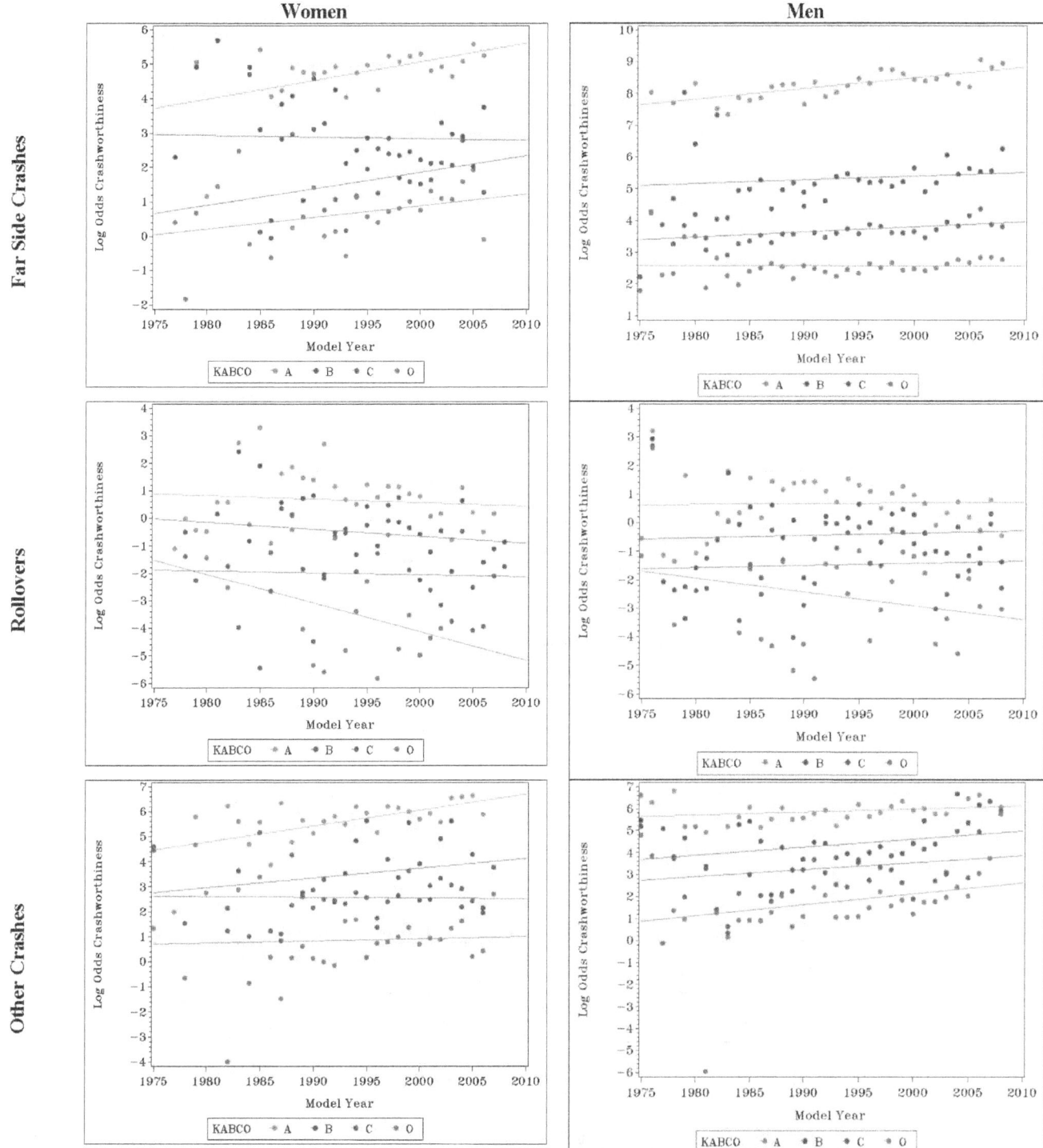

9.2 Parameter Estimates for the Crash Avoidance Model

Table A-1: The Crash Rate for Frontal Crashes of Cars During the Calendar Year of Their Model Year From the Final Crash Avoidance Model

Crash Rate	P-Value
0.14999	<0.0001

Table A-2: Multiplicative Effects on the Crash Rate in the Final Crash Avoidance Model

Effect	Level	Multiplicative Effect on Crash Rate	
		Estimate	Probability of No Effect
Calendar Year		0.9730	<0.0001
Crash Type	Rollover	0.0094	<0.0001
	Side	0.7018	<0.0001
	Other	0.6657	<0.0001
Vehicle Age	NA	1.0988	<0.0001
Vehicle Age^2	NA	0.9953	<0.0001
Vehicle Type	LTV	0.9379	0.0059
CrashType *CalendarYear	Rollover	0.9612	<0.0001
	Side	0.9930	0.2693
	Other	1.0016	0.8043
CrashType *VehicleAge	Rollover	1.0287	0.0112
	Side	0.9736	0.0103
	Other	0.9194	<0.0001
CrashType *VehicleAge^2	Rollover	0.9987	0.0153
	Side	1.0006	0.2413
	Other	1.0020	<0.0001
CrashType *VehicleType(LTV)	Rollover	2.1927	2.1927
	Side	0.9572	0.9572
	Other	1.2445	1.2445
Dispersion	NA	1.0524	<0.0001

9.3 Parameter Estimates for the Crashworthiness Model

Although 1985 constitutes the earliest model year in the training data to which the model was fit, we use model year 2000 in our baseline group as we are primarily interested in recent vehicle improvements. That is, the estimate for the injury odds of a given KABCO level in Table A-3 is equal to the product of the estimates for the parameters Intercept and Model Year for the crashworthiness model of the KABCO level.

Table A-3: The Odds of Injury for Unrestrained 25- to 65-Year-Old Women in Frontal Crashes of Model Year 2000 Cars in the Final Crashworthiness Model

KABCO	Injury Odds		The Multiplicative Effect of a One Year Increase in Model Year on the Injury Odds	
	Estimate	Probability of No Effect	Estimate	Probability of No Effect
O	1.2	<0.0001	1.0351	<0.0001
C	3.0	<0.0001	1.0346	<0.0001
B	9.3	<0.0001	1.0292	0.00
A	83.2	<0.0001	1.0046	0.59

Table A-4 presents the multiplicative effects from the model.

Table A-4: Multiplicative Effects on the Odds of Injury in the Final Crashworthiness Model

Effect	Level	KABCO	Multiplicative Effect on Injury Odds		The Multiplicative Effect of a One Year Increase in Model Year on the Injury Odds	
			Estimate	Probability of No Effect	Estimate	Probability of No Effect
Occupant Age	<14	O	1.19	0.07	4.09	1.63
		C	1.12	0.27	4.09	1.63
		B	1.31	0.17	4.09	1.63
		A	1.81	<0.0001	4.09	1.63
	14-24	O	1.22	<0.0001	1.03	0.00
		C	1.13	0.01	1.03	0.00
		B	1.33	<0.0001	1.01	<0.0001
		A	1.53	<0.0001	0.99	0.00
	66+	O	0.93	0.06	1.03	0.12
		C	0.72	<0.0001	1.03	0.41
		B	0.66	<0.0001	1.03	0.75
		A	0.27	<0.0001	1.00	0.18
Crash Type	Rollover	O	0.12	<0.0001	0.99	0.05
		C	0.19	<0.0001	0.99	0.01
		B	0.18	<0.0001	0.98	<0.0001
		A	0.06	<0.0001	0.99	0.03
	Near Side	O	1.04	0.73	1.04	0.18
		C	1.09	0.40	1.04	0.62
		B	0.89	0.44	1.03	0.93
		A	0.52	<0.0001	1.02	0.11
	Far Side	O	1.32	0.01	1.04	0.50
		C	1.29	0.02	1.04	0.42
		B	1.00	0.82	1.04	0.17
		A	0.96	0.56	1.01	0.16
	Other	O	1.53	0.01	1.04	0.33
		C	2.88	<0.0001	1.05	0.29
		B	2.35	<0.0001	1.04	0.27
		A	1.54	<0.0001	1.03	<0.0001
Driver Alcohol	alc involved	O	0.53	<0.0001	1.02	0.02
		C	0.41	<0.0001	1.02	0.11
		B	0.39	<0.0001	1.02	0.12
		A	0.28	<0.0001	0.99	0.00
	no driver	O	1.89	0.38	0.96	0.08
		C	3.01	0.44	0.94	0.17
		B	1.82	0.22	0.96	0.09
		A	1.38	0.62	0.96	0.37
Gender	Male	O	1.61	<0.0001	0.00	<0.0001
		C	1.35	<0.0001	0.00	<0.0001
		B	1.39	<0.0001	0.00	<0.0001
		A	0.92	<0.0001	0.00	<0.0001
Restraint Use	Restrained	O	4.09	<0.0001	0.00	<0.0001
		C	5.35	<0.0001	0.00	<0.0001
		B	5.50	<0.0001	0.00	<0.0001
		A	11.49	<0.0001	0.00	<0.0001
Vehicle Type	LTV	O	1.31	<0.0001	0.00	<0.0001

Effect	Level	KABCO	Multiplicative Effect on Injury Odds		The Multiplicative Effect of a One Year Increase in Model Year on the Injury Odds	
			Estimate	Probability of No Effect	Estimate	Probability of No Effect
		C	1.19	<0.0001	0.00	<0.0001
		B	1.09	0.14	0.00	<0.0001
		A	1.15	0.00	0.00	<0.0001
Crash Type * Restraint Use	Rollover *Restrained	O	1.25	0.51	1.06	0.22
		C	0.84	0.15	1.06	0.07
		B	0.97	0.40	1.05	0.02
		A	0.95	0.41	0.98	0.05
	Near Side *Restrained	O	0.80	0.03	1.03	0.59
		C	0.79	0.03	1.04	0.78
		B	0.80	0.13	1.04	0.51
		A	0.85	0.03	0.99	0.16
	Far Side *Restrained	O	0.82	0.06	1.04	0.90
		C	0.94	0.69	1.03	0.84
		B	1.21	0.19	1.02	0.42
		A	1.26	<0.0001	0.99	0.11
	Other *Restrained	O	0.44	<0.0001	1.04	0.52
		C	0.52	<0.0001	1.04	0.39
		B	0.56	0.00	1.04	0.51
		A	1.41	<0.0001	1.00	0.54
Crash Type * Vehicle Type	Rollover *LTV	O	0.97	0.61	1.03	0.61
		C	0.85	0.48	1.03	0.69
		B	0.89	0.40	1.03	0.70
		A	1.22	0.04	0.99	0.15
	Near Side *LTV	O	1.10	0.06	1.03	0.39
		C	1.17	0.01	1.03	0.38
		B	1.54	<0.0001	1.02	0.07
		A	1.94	<0.0001	1.01	0.78
	Far Side *LTV	O	1.14	0.02	1.03	0.69
		C	1.19	0.01	1.03	0.65
		B	1.39	0.00	1.03	0.87
		A	1.59	<0.0001	1.01	0.03
	Other *LTV	O	1.08	0.09	1.04	0.73
		C	1.05	0.28	1.04	0.62
		B	1.31	0.00	1.03	0.78
		A	1.85	<0.0001	1.00	0.72
Driver Alcohol * Restraint Use	alc involved *Restrained	O	1.31	<0.0001	NA	NA
		C	1.11	0.10	NA	NA
		B	1.09	0.43	NA	NA
		A	0.61	<0.0001	NA	NA
	no driver *Restrained	O	0.73	0.42	NA	NA
		C	0.34	0.08	NA	NA
		B	0.52	0.10	NA	NA
		A	0.15	<0.0001	NA	NA

www.ingramcontent.com/pod-product-compliance
Lightning Source LLC
Chambersburg PA
CBHW081836170526
45167CB00007B/2833